Nehal Salahuddin

Polymer-Clay Nanocomposites for Drug Delivery Systems

Nehal Salahuddin

Polymer-Clay Nanocomposites for Drug Delivery Systems

How nanoclays can heal burn infection timely and effectively

LAP LAMBERT Academic Publishing

Impressum / Imprint

Bibliografische Information der Deutschen Nationalbibliothek: Die Deutsche Nationalbibliothek verzeichnet diese Publikation in der Deutschen Nationalbibliografie; detaillierte bibliografische Daten sind im Internet über http://dnb.d-nb.de abrufbar.

Alle in diesem Buch genannten Marken und Produktnamen unterliegen warenzeichen-, marken- oder patentrechtlichem Schutz bzw. sind Warenzeichen oder eingetragene Warenzeichen der jeweiligen Inhaber. Die Wiedergabe von Marken, Produktnamen, Gebrauchsnamen, Handelsnamen, Warenbezeichnungen u.s.w. in diesem Werk berechtigt auch ohne besondere Kennzeichnung nicht zu der Annahme, dass solche Namen im Sinne der Warenzeichen- und Markenschutzgesetzgebung als frei zu betrachten wären und daher von jedermann benutzt werden dürften.

Bibliographic information published by the Deutsche Nationalbibliothek: The Deutsche Nationalbibliothek lists this publication in the Deutsche Nationalbibliografie; detailed bibliographic data are available in the Internet at http://dnb.d-nb.de.

Any brand names and product names mentioned in this book are subject to trademark, brand or patent protection and are trademarks or registered trademarks of their respective holders. The use of brand names, product names, common names, trade names, product descriptions etc. even without a particular marking in this work is in no way to be construed to mean that such names may be regarded as unrestricted in respect of trademark and brand protection legislation and could thus be used by anyone.

Coverbild / Cover image: www.ingimage.com

Verlag / Publisher:
LAP LAMBERT Academic Publishing
ist ein Imprint der / is a trademark of
OmniScriptum GmbH & Co. KG
Heinrich-Böcking-Str. 6-8, 66121 Saarbrücken, Deutschland / Germany
Email: info@lap-publishing.com

Herstellung: siehe letzte Seite /
Printed at: see last page
ISBN: 978-3-659-69228-4

Copyright © 2015 OmniScriptum GmbH & Co. KG
Alle Rechte vorbehalten. / All rights reserved. Saarbrücken 2015

Polymer-Clay Nanocomposites for Drug Delivery Systems
How nanoclays can heal burn infection timely and effectively

Nehal A. Salahuddin
Department of Chemistry, Polymer Research Group, Faculty of Science,
Tanta University, Tanta 31527, Egypt

Abstract

Polymer-montmorillonite nanocomposites are an important class of materials that can be utilized in applications such as dental and engineering materials, drug delivery, tissue engineering, and antimicrobial treatment. The incorporation of nanoclays into a polymer matrix can provide unique properties and can also improve properties such as mechanical strength and thermal behavior. This chapter covers the background, definitions, and some potential applications of polymer-montmorillonite nanocomposites as drug carrier and antimicrobial materials. It covers some types of polymers used in the preparation of montmorillonite nanocomposites. The morphology of the polymer nanocomposites are discussed along with the release results as well as the antimicrobial activity of the polymers themselves. The limited data available on the *in vivo* study of the nanocomposites is also presented.

Contents

Nehal A. Salahuddin

Department of Chemistry, Faculty of science, Tanta University, Tanta, Egypt.

e-mail: salahuddin.nehal@yahoo.com

       :nehal.attaf@science.tanta.edu.eg

1. Introduction

Polymer-montmorillonite nanocomposites are an important class of materials that can be utilized in applications such as drug delivery, tissue engineering, dental and engineering materials. The incorporation of nanocaly into a polymer matrix can provide unique properties and also improve properties such as thermal properties and mechanical strength. An advantage of using hydrogel nanocomposites in biomedical applications stems due to their similarity to soft tissue as well as expected biocompatibility. Overall, the safety concern of polymer nanocomposites involves the biocompatibility of the system when implanted in the body. As such, it is important to test the effect of the nanocomposite on the body before it is utilized in any biomedical application. A limited number of polymer nanocomposites have been studied for safety in biomedical applications. One of the most significant studies involves a clay/hydrogel composite that can be used for cell cultivation (Haraguchi, Takehisa et al. 2006, Zheng, Luan et al. 2007). In these studies, cell cultivation on the surface of a nanocomposite was shown to be successful for three lines of cells including HepG2 human hepatoma cells, human dermal fibroblasts, and human umbilical vein endothelial cells. In another publication a nanocomposite showed no cytotoxic effects and no inhibition of osteoblast (bone cell) proliferation (Giordano, Sanginario et al. 2006). The principal objective of the

2

present chapter is meant to overview some of the efforts done on the synthesis of polymer-montmorillonite nanocomposites as drug delivery and antimicrobial materials. This chapter reviews the background concerning montmorillonite and polymer nanocomposites along with the applications of these materials as drug delivery system and antimicrobial materials. Comparing the potential and reality it is pretty obvious that though these deliver systems have a huge potential as antimicrobial packaging to inhibit microbial growth and extend the shelf life of foods, we still need to do extensive research to make it a reality. The research should move from just showing the potential of the delivery systems to actually proving their efficacy in experimental animals.

2. Hierarchical structure and characteristics of montmorillonite

Montmorillonite (MMT) is a hydrous–alumina silicate mineral whose lamellae are constructed from an octahedral alumina sheet sandwiched between two tetrahedral silica sheets (Fig.1a). A net negative charge on the layers surface causes them to attach by sorbed Na^+ cations. The layers of MMT are not stacked regularly enough to form true crystals but are grouped in small numbers forming stacks, inside which the degree of organization is a parallelism sufficient to result in selective 001 reflection.

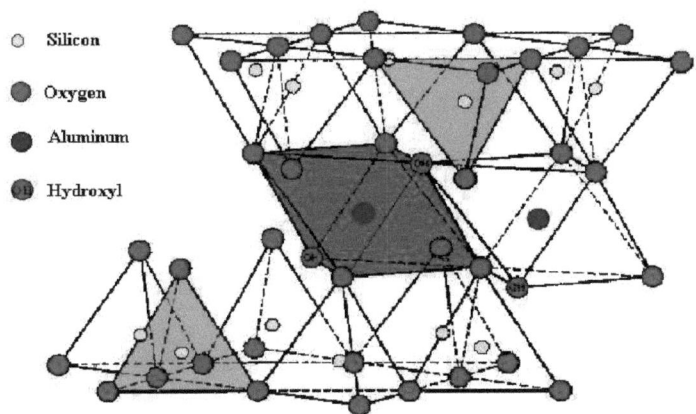

○ Silicon

● Oxygen

● Aluminum

● Hydroxyl

Fig.1a. Idealised structure for montmorillonite lamella.

The various organization levels in MMT (Akelah, Salahuddin et al. 1994) are: a) primary particles (*ca.* 10 nm) consist of stacks of parallel elementary sheets (lamellae) with an average of 10 sheets per particle. B) Microaggregates (*ca.* few hundreds nm) formed by the association of several primary particles which are nearly parallel and joined together. C) Aggregates (0.10-10 μm) comprise a large number of primary particles and microaggregates (Fig. 1b).

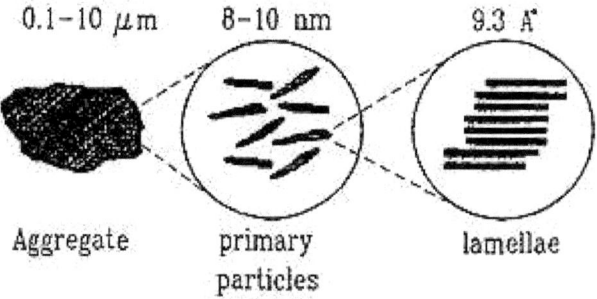

Fig.1b.Hierarchical structure of montmorillonite.

The irregular character of the distribution of the charges in MMT weakens the attraction between the interlayers enough for water to force them apart and enter into the spaces, making the whole layer accessible to hydration and cations intercalation. The swelling and plasticizing of the MMT are due to increasing interparticle spaces rather than to the expansion of individual primary particles. Thus, MMT undergoes two interdependent processes during intercalation (Mering 1946) swelling of the interlamellar spacing and disaggregation. This multifarious structure of the mineral furnishes abundant surface area (780 m²/g) for the host molecules to interact by intercalation and otherwise. In clay modification, mostly interlayer cations (Na^+) are replaced by organic bulky ammonium or phosphonium cations (Yano, Usuki et al. 1993, Akelah and Moet 1994) (Fig.2). This leads to an increase in interlayer spacing and a decrease in clay layer-layer attraction. Therefore, the MMT is suggested to be a good delivery carrier for the hydrophilic drugs. It can absorb excess water from feces and thus act as anti-diarrheic. MMT can also provide mucoadhesive capability for the nanoparticles to cross the gastrointestinal barrier (Dong and Feng 2005); (Feng, Mei et al. 2009). MMT has been proved to be nontoxic by hematological, biochemical and histopathological analyses in rat models (Lee, Kuo et al. 2005).

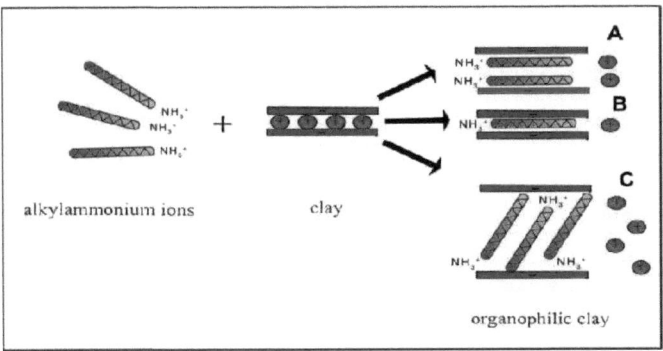

Fig. 2.The cation-exchange process between alkylammonium ions and cations initially intercalated between the clay layers. (A) lateral bilayer; (B) lateral mono layer; (C) paraffin type monolayer arrangement of alkylammonium ions between the montmorillonite layers.

3.General methods of preparation polymer-clay nanocomposites

The preparative methods are divided into three main groups according to the starting materials and processing techniques:

3.1.*In situ* polymerization method

This process is conventionally used to synthesize a thermoset-clay nanocomposite. The organoclay (surface treated clay) is swollen in the monomer. Then, the reaction is initiated by addition of a curing agent or initiator (Salahuddin, Moet et al. 2002, Salahuddin and Shehata 2002, Salahuddin, Ayad et al. 2008). Figure 3a shows a schematic diagram of *in situ* polymerization method.

The swelling phase, the high surface energy of the clay attracts polar monomer molecules so that they diffuse between the clay platelets. Later, the polymerization reaction lowers the overall polarity of the intercalated molecules and displaces the thermodynamic equilibrium in such a way that more polar molecules are driven in between the clay layers and delaminate the clay eventually.

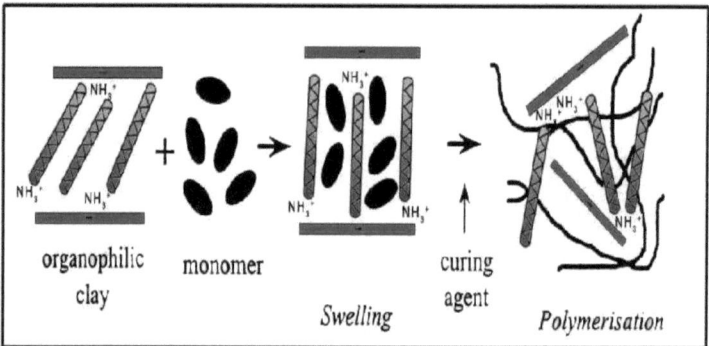

Fig. 3a. A schematic diagram of *in situ* polymerization method.

3.2. Melt intercalation method

Melt intercalation is used to synthesize nanocomposites based on thermoplastics. Molten thermoplastic is directly blended with organoclay in an extruder in order to optimize the polymer-clay interactions. The mixture is then heated and molded into any desired shape and the nanocomposite is formed (Vaia, Ishii et al. 1993).

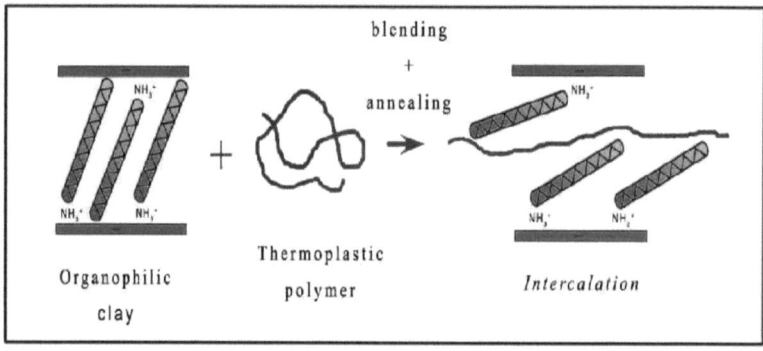

Fig.3b. A schematic diagram of melt process.

The melt intercalation process is becoming increasingly popular because of its simplicity for application in industry. A schematic diagram of this process is shown in Fig. 3b.

3.3. Solvent Method

In this technique, clay or organoclay is swollen in a solvent. Polymer dissolved in suitable solvent is added to the swollen clay mineral (Salahuddin, El-Barbary et al. 2009 a,b, Salahuddin and Abdeen 2012 a,b, Salahuddin, Badr et al. 2012, Salahuddin, Kenawy et al. 2012, Salahuddin, El-Barbary et al. 2014). The polymer intercalates between the clay platelets and the solvent is then removed by evaporation under vacuum. This approach is not practical for industrial use due to the problems associated with removing a large quantity of solvent. Nevertheless, this process offers the possibilities of synthesizing intercalated nanocomposites based on polymers with low or even no polarity. A schematic diagram of this process is shown in Fig.3c.

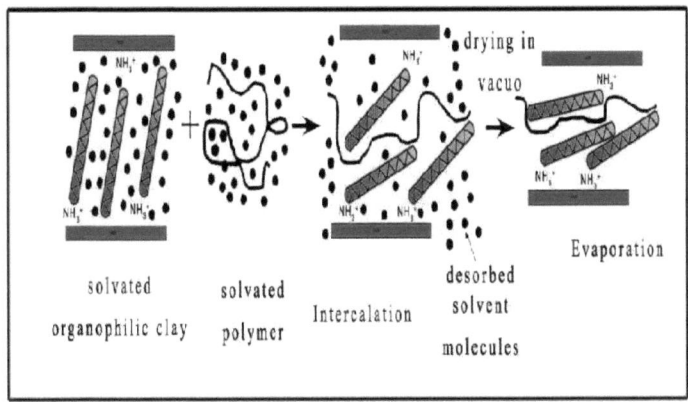

Fig.3c. A schematic diagram of solvent process.

In this chapter a number of polymer-montmorillonite nanocomposites have been prepared using this method as drug delivery system and biological active materials. Preparation of polyoxy propylene–MMT nanocomposites by intercalation of polyoxypropylene diamine using three different molecular masses (D_{230}, D_{400}, D_{2000}) into MMT galleries was carried out by a cationic-exchange process followed by Intercalation of drugs into the poly oxypropylene–MMT nanocomposites (D_{230}–MMT, D_{400}–MMT, D_{2000}–MMT)(Salahuddin, Kenawy et al. 2012). The role of polyoxypropylene hydrochloride in the modified clay mineral is to lower the surface energy of the inorganic host and improve the wetting characteristics with the (drug) organic compounds. Additionally, the polyoxypropylene diamine could provide functional group that can react with the aldehydes and esters (Salahuddin, Badr et al. 2012 a,b). The bioactive agents p-hydroxymethylbenzoate, 2,4-dihydroxymethyl benzoate, methyl salicylate, vanillin (4-hydroxy-3-methoxy benzaldehyde) and 5-formylamino salicylic acid microbicides were reacted with $D_{230-2000}$-MMT nanocomposites.

New microbicidal polyamides were prepared by the reaction of 5-phenyl-1,3,4,-oxadiazole-2-thiol, 5-phenyl-1,3,4-oxadiazole-2-amine, and 5-(4-chlorophenyl)-1,3,4-thiadiazole-2-thiol with ethyl chloroformate followed by polycondensation with polyoxy-propylenetriamine (Jeffamine T_{403}) (Salahuddin, El-Barbary et al. 2014). The polyamides were modified to yield amine hydrochloride. The intercalation of polyamides into MMT was achieved through an ion exchange process between sodium cations in MMT and amine hydrochloride in the polyamides. A schematic diagram of this preparation is shown in scheme1.

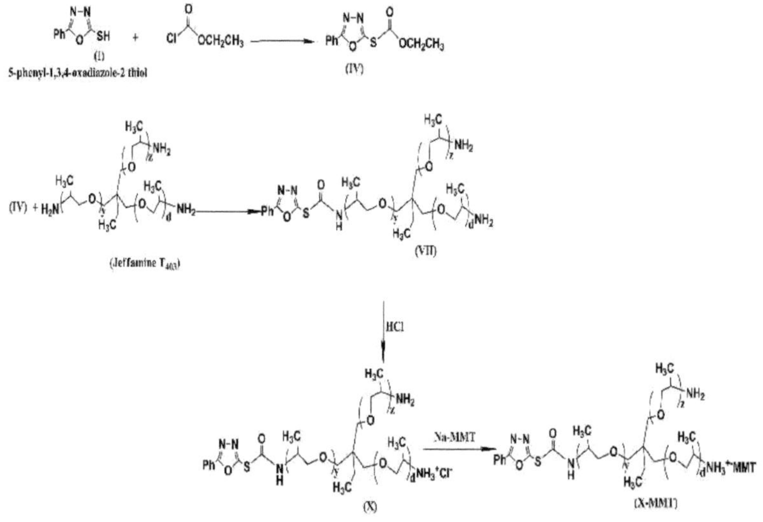

Scheme 1. A schematic diagram of microbicidal polyamides-MMT nanocomposites.

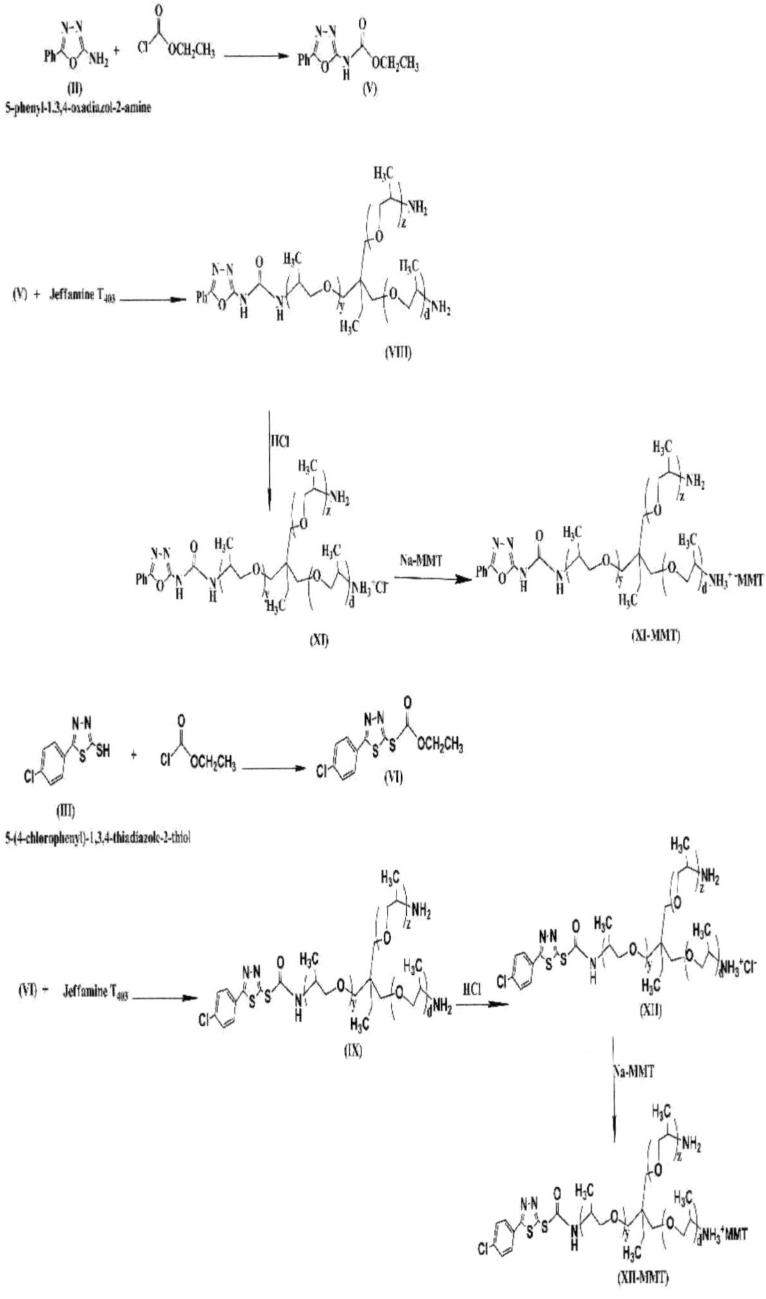

Scheme 1. continued.

4. Nanocomposites as drug delivery system

Drug delivery systems have been of great interest for the past few decades to realize the effective and controlled drug delivery and minimize the side effects in the field of pharmaceutics. Oral controlled drug delivery system is an essential part of the development of new medicines. The carriers used for control drug release were mainly biodegradable polymers (Langer, Kathryn et al. 1999) and porous inorganic matrix (Aguzzi, Cerezo et al. 2007, Suresh, Borkar et al. 2010). In recent years, drug intercalated smectite, especially MMT pharmaceutical grade clay mineral has attracted great interest for researchers (Joshi, Patel et al. 2009 a,b).

4.1. Clay minerals as drug delivery system

MMT is utilized as a sustained release carrier for various therapeutic molecules, such as 5 Fluorouracil (Lin, Lee et al. 2002), sertraline (Nunes, Vaz et al. 2007), vitamin B(Joshi, Patel et al. 2009 a,b), promethazine chloride (Seki and Kadir 2006), ibuprofen (Zheng, Luan et al. 2007), promethazine chloride and buformin hydrochloride (Fejer, Kata et al. 2001), timolol maleate (TM) (Joshi, Kevadiya et al. 2009), Procainamide (Kevadiya, Joshi et al. 2010) and buspirone hydrochloride (Joshi, Kevadiya et al. 2010). Smectite clay minerals intercalated by drug molecules exhibit novel physical and chemical properties (Joshi, Patel et al. 2009 a, Kevadiya, Joshi et al. 2010). However, in spite of the beneficial effect of clay, there are some inherent drawbacks associated with the use of clay for drug delivery. Under physiological conditions clay dispersions are unstable and tend to flocculate and precipitate in ion containing solutions because of high salt concentration and the presence of polyelectrolytes such as proteins. Stability of dispersion is an important requirement for drug carriers because it plays a determining role with regard to adsorption and bioavailability. Furthermore, the ability of clay particles to adsorb negatively charged or neutral drugs are low, restricting their application as carriers of negatively charged or neutral drugs (Aguzzi, Cerezo et al. 2007). In this regard, it is believed that the synthesis of polymer nanocomposites would alleviate this disadvantage by exploiting the properties of clay and polymer in such a way that the behavior of the clay mineral is modified.

4.2.Polymer/clay nanocomposites as drug delivery system

Nanocomposites based on poly (ethylene-co-vinyl acetate) (EVAc) and three different organo-silicates (one montmorillonite and two synthetic micas) have been developed (Cypes and Saltzman 2003) providing slow release of dexamethasone (a corticosteroid agent widely used to reduce inflammatory diseases), as well as enhanced mechanical properties (Young's modulus) in comparison with the free polymer. In vitro release studies of differently charged drugs from N-isopropylacrylamide-montmorillonite (Lee and Fu 2003) showed that the amount released was lower when the drug and hydrogel were oppositely charged. Conversely when the drug and hydrogel showing the same sign of charge, the amount released was higher. The amount of the organic agent intercalated into montmorillonite affect on the physical properties and drug release behavior from N-isopropylacrylamide-montmorillonite nanocomposites (Lee and Jou 2004). Yuan, Shah et al. 2010 studied the controlled release of drug from a chitosan–clay nanocomposite drug carrier, in contrast to pure chitosan. The release is controlled by electrostatic interaction between the positive charge of Doxorubicin Hydrochloride (DOX) and negatively charged sites in the clay mineral. The factors governing the drug release profile include swelling behavior and drug–carrier interactions. The drug release behavior is influenced by pH and the chitosan/clay ratio. Drug release occurs by degradation of the nanocomposite particle carrier to its individual components or nanostructures of different composition. Pongjanyakul, Priprem et al. 2005, showed that rheological properties (such as viscosity and flow behavior) of sodium alginate or poloxamer 407 gels were improved by incorporation of magnesium aluminium silicate (MAS), while less influence was observed for hydroxypropyl methylcellulose (HPMC) gels. Drug release of these composite materials was significantly decreased in comparison with the pure polymeric gels, due to both clay-polymer interactions, and drug adsorption onto clay particles. Magnesium aluminium silicate has also been applied to improve physical properties in alginate beads and to prepare novel composite films as coating material for modifying drug release. Takahashi, Yamada et al. 2005 have developed a nanocomposite system by combining Laponite (a synthetic hectorite-type clay mineral) with a block copolymer containing poly (ethylene glycol) and polyamine segments. These nanoparticles were capable of sustained

11

release of pyrene (hydrophobic model drug molecule) over a period of 20 days. Flocculation-resistive properties of these hybrid particles at high ionic strength have also been reported, whereas clay mineral dispersions normally tend to coagulate in salt solutions, as has been widely documented (Lukham and Rossi 1999, Ma and Pierre 1999). (Dong and Feng 2005) have designed a drug delivery system, consisting of poly (D,L-lactide-co-glycolide)- montmorillonite nanocomposites, for oral delivery of paclitaxel. The clay mineral did not affect drug release from paclitaxel-loaded particles but it promoted the cellular uptake efficiency of the PLGA nanoparticles in human colonic adeno carcinoma cells (Caco-2) and in human colon adenocarcinoma grade II cell line (HT-29) cells. Biologically active compounds including 4-amino-6-methyl-3-thioxo-3,4dihydro-2H-[1,2,4]triazin-5-one, 4-amino-6-methyl-3-thioxo-3,4-dihydro-2H [1,2,4] triazin-3,5-dithione, 4-amino-6-(4-methoxy styryl) -3-thioxo-3,4dihydro-2H-[1,2,4]triazin-5-one and 4-amino-6-styryl-3-thioxo-3,4-dihydro2H-[1,2,4] triazin-5-one, have been reacted with polymethylmethacrylate (PMMA)-MMT intercalates and ion exchanged with sodium montmorillonite (MMT) (Salahuddin, El-Barbary et al. 2009 b). The release of biologically active compounds intercalated layered silicates is controllable and these materials have a great potential as a delivery host in the pharmaceutical field. In another publication (Salahuddin, El-Barbary et al. 2009 a), biologically active compounds including 4-amino-6-substituted-3-thioxo-3,4-dihydro-2H-5-oxo(or-thioxo)1,2,4 triazines were reacted with cross-linked PMMA and compared with Triazine-MMT. The release of triazine compounds is more controllable in intercalated layered silicates. Quaternized chitosan/montmorillonite nanocomposites (Wang, Du et al. 2008) were modified to prepare the nanoparticles, whose drug-controlled release behaviors were evaluated. Compared to pure nanoparticles, certain MMT loadings on quaternized chitosan enhanced the drug encapsulation efficiency of the nanoparticles and slowed the drug release from the nanocomposites. MMT- propranolol hydrochloride–poly lactic-co-glycolic acid nanocomposites are supposed to be better oral controlled drug delivery system, for a highly hydrophilic low molecular weight antihypertensive drug propranolol to minimize the drug dosing frequency and hence improving the patient compliance (Seema, 2013).

4.3. Site-specific polymer/montmorillonite nanocomposite drug delivery systems (morphology vs release)

Site-specific drug delivery to the colon has attracted increasing attention, both for therapy of colon-related diseases and systemic drug delivery. For this purpose, it is necessary to incorporate the drug in a formulation able to minimize premature release in the upper part of the gastrointestinal tract, and then to optimize drug release into the colon. Polyoxypropylene MMT/Theophylline and polyoxypropylene-MMT/Ibuprofen using montmorillonite modified by three kinds of polyoxypropylene ($D_{230-2000}$) were prepared in order to increase and attain extended release in aqueous solutions and direct drug release into the colon (Salahuddin, Kenawy et al. 2012). X-ray diffraction pattern displayed in Fig. 4 indicates a regular lattice spacing of 14.7 Å for D_{230}-MMT, 16.3Å for D_{400}-MMT and 18.4 Å for D_{2000}-MMT with an expansion of the clay galleries by 5.4Å, 7Å and 9.1Å, respectively. After D_{230}MMT was incorporated with Ibuprofen, the basal spacing characteristic to D_{230}MMT was substituted by a broader peak. However, D_{400}-MMT / Ibuprofen exhibit a diffraction shoulder. This indicated that a substantial part of clay mineral is only exfoliated. In case of D_{2000}-MMT / Ibuprofen, the peak characteristic of D_{2000}-MMT was virtually disappeared with the appearance of two broad peaks at $2\theta= 2.8, 6°$ indicating the formation of an order-intercalated structure. However, in D_{230}-MMT /Theophylline a sharp peak is obtained. The high relative intensity of the 001 peak is due to a more narrow distribution of the interlamellar spacing. It is worth noting that the basal spacing of Na-MMT was expanded from 9.6 to 13.6 A˚ in Na-MMT/ibuprofen; this indicated the intercalation of ibuprofen between the layers of Na-MMT.

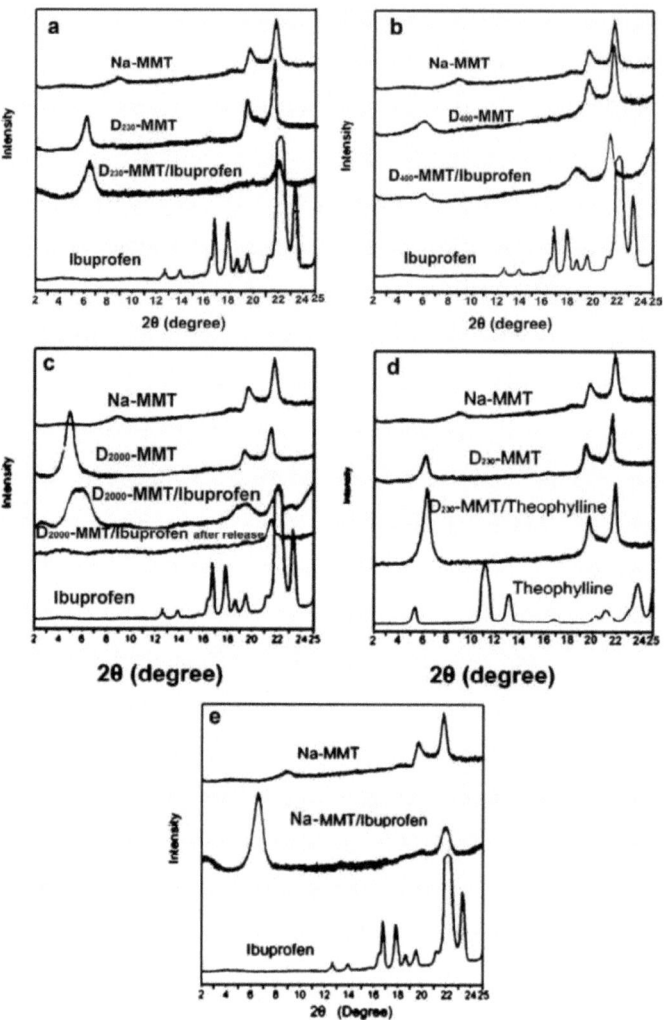

Fig. 4. XRD patterns of Na-MMT, ibuprofen, theophylline , polyoxypropylene–MMT, Na-MMT / drug and polyoxypropylene–MMT / drug.

TEM of ultramicrotome sections prepared from D_{230}–MMT/ibuprofen (Fig. 5(a)), D_{400}–MMT/ibuprofen (Fig. 5(b)), and D_{2000}–MMT/ibuprofen (Fig. 5(c)) specimens embedded in epoxy resin displayed typical portraits of the nanomineral domains. D_{230}– MMT/ibuprofen showed the existence of an intercalated structure. In D_{400}–MMT/ibuprofen, the domains appeared to be

14

flocculated. However, in D_{2000}–MMT/ibuprofen, TEM revealed the presence of ordered multiplets with an average size of 50 nm.

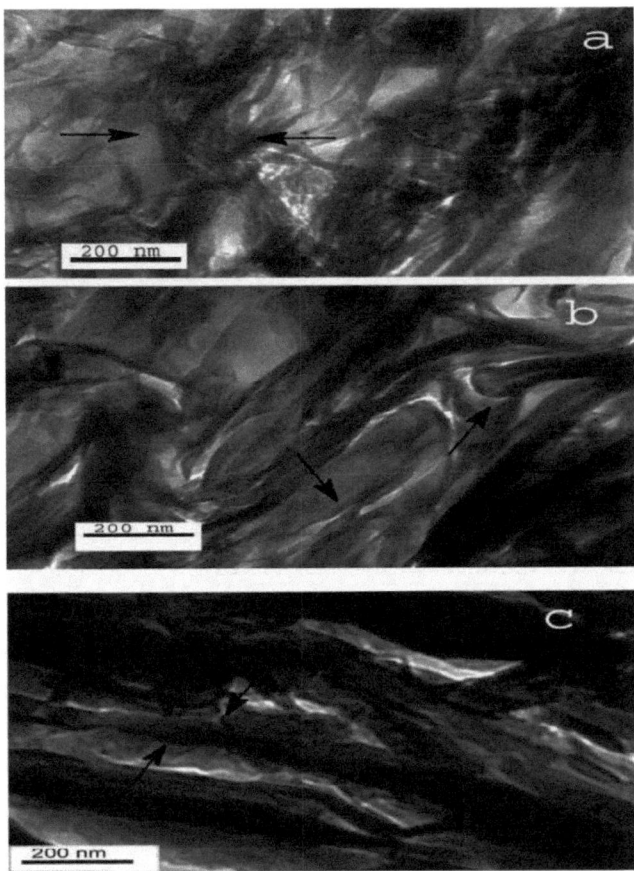

Fig. 5. TEM micrographs of (a) D_{230}–MMT/ibuprofen, (b) D_{400}– MMT/ ibuprofen, and (c) D_{2000}–MMT/ibuprofen.

An interesting observation was that no release was observed at pH 5.4. This study suggested that the carrier would not release the drug in gastrointestinal fluid after oral administration. It was reported that ibuprofen (Zheng, Luan et al. 2007) could be embedded into Na-MMT at pH 11. The *in vitro* release of ibuprofen from Na-MMT was noticed in gastric acid fluid (pH 1.2) and simulated intestinal fluid (pH 7.4). This may have been due to the intercalation of ibuprofen in the form of sodium salt that favored the solubility and release of the drug in gastric acid fluid. It is worth mentioning that the intercalation of polyoxypropylene into the clay built a strong crosslinking structure because of the negatively charged clay and positively charged -NH_3^+ group of the polymer. This influenced the swelling behavior of the support and consequently, the diffusion of the drug through the bulk entity.

With increasing molecular mass, a significant decrease in the swellability of the carrier led to a decrease in the burst release of the drug. In alkaline medium, the hydrolysis of the amide linkages took place through the attack of the nucleophile on the electron deficient carbonyl carbon, and the high solubility of the drug in the alkaline medium led to their release. It has been suggested that drug diffusion was hindered by the three dimensional (house-of-cards) network formed by clay mineral via edge face interactions. The swellability percentage for Na-MMT and its derivatives ($D_{230-2000}$-MMT) in pH=7.8 was determined to be Na-MMT =232.8, D_{230}-MMT=262.4, D_{400}-MMT=23.6 and D_{2000}-MMT=1.5%).

In pH 7.8, the hydrolysis of amide linkages takes place through the attack of the nucleophile on the electron deficient carbonyl carbon and the high solubility of drug in alkaline medium favor the release of the drugs. The release profile in Fig. 6 shows that the maximum amount of Ibuprofen released is almost 63% from D_{230}-MMT/Ibuprofen, 86% from D_{400}-MMT/Ibuprofen, 53% from D_{2000}MMT/ Ibuprofen, 30 % from Na-MMT/Ibuprofen (Fig. 6a). However, the maximum amount of Theophylline released is 55%. Comparison between the release rate of Ibuprofen from D_{230}-MMT/Ibuprofen (drug content 10.77%), D_{400}-MMT/Ibuprofen (drug content 5.56%), D_{2000}-MMT/Ibuprofen (drug content 1.82) indicated that the release from partially flocculated structure with low loading drug (D_{400}-MMT/Ibuprofen) is higher than that from intercalated structure with a high loading drug D_{230}-MMT/Ibuprofen. The amount released of Ibuprofen (drug content 10.77%) was higher than amount of Theophylline (drug content 16.08%) released from the same support (Fig.6b). These results indicate that the morphology of the support, loading and the type of interaction between the drugs and support affect on the rate of release. It was reported that

16

the rate of drug diffusion out of the matrix is controlled by the rigidity of the layers and the diffusion path length(Ambrogi 2001). In addition, the interactions between the drug and MMT should successfully prolong the action of drug (Maheshwari, Sharma et al. 1988). In contrast to the release of other drugs from intercalated LDH materials reported in the literature (Suzuki, Nakamura et al. 2001) the release of drugs from organo-clay was not complete within 80 min. The cause of the partial release was attributed to the possibility that the drug molecules are deeply embedded in the organo-clay and complete release is very slow.

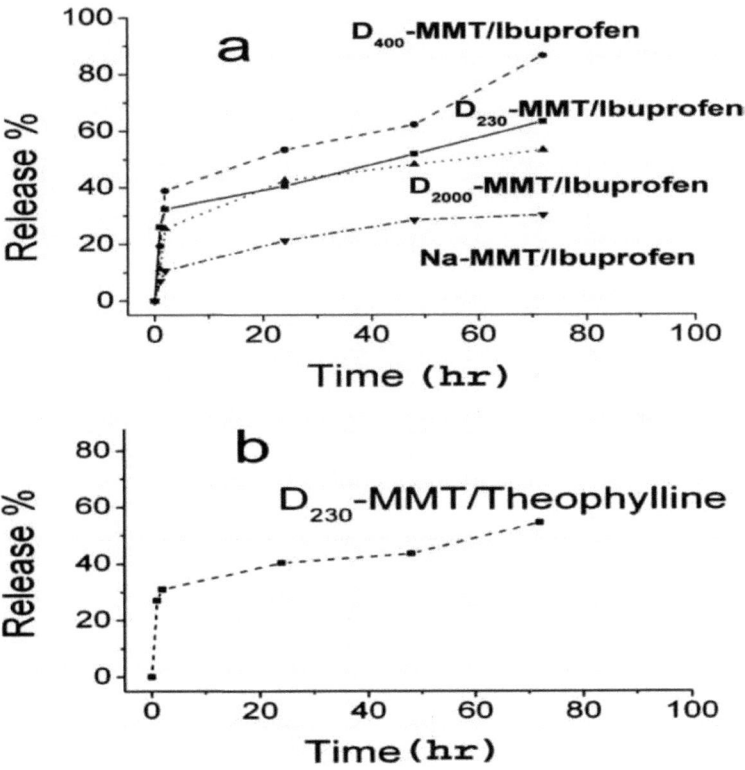

Fig. 6. *In vitro* release of Ibuprofin from a) Na-MMT/Ibuprofen and $D_{230-2000}$ MMT/ Ibuprofen nanocomposites; b) Theophylline from D_{230} – MMT/ Theophylline in phosphate buffer solution (pH = 7.8) at 37°C.

It is worthnoting that a broad peak (Fig. (4)) was observed for D_{2000}-MMT/ Ibuprofen after drug release at lower d-spacing (2Θ =4°) due to the release of drug. TEM micrographs of D_{2000}-MMT / Ibuprofen at identical magnifications before and after drug release at the selected pH of 7.4 are presented in Fig. (7). A monodispersion with an average size of 20 nm is observed in Fig. (7a) implies that the size of the nanoparticle after drug release was significantly increased to 70 nm (Fig. 7b). The reason for the increase in size after drug release is believed to be a consequence of the detachment or separation of drug which lead to aggregation of the clay particles.

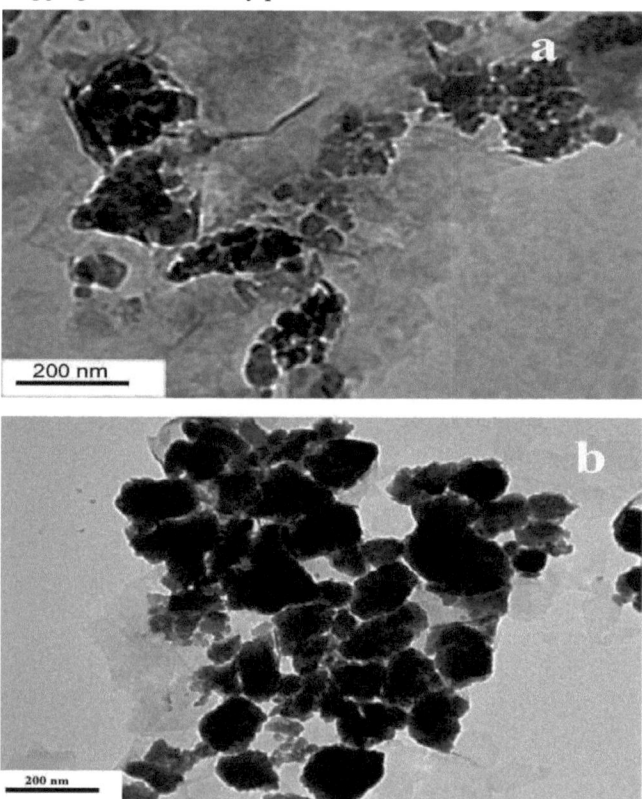

Fig. 7. TEM micrograph of D_{2000}-MMT/ Ibuprofen a) before and b) after release

5. Nanocomposites with antimicrobial activity

Infection by bacteria is quite problematic in many fields, including food packaging and hospital furniture. These microorganisms are indeed pathogen and thus responsible for many diseases (Takai, Ohtsuka et al. 2002). In order to get rid of them biocides, i.e. chemicals that inhibit the growth of microorganisms, have been made available commercially, as alcohols (Morton, 1983), biguanides (Biase, 1980) and halogen releasing agents (Bloomfield, 1996). Cationic agents of the quaternary ammonium type have been shown to be potential antiseptic and disinfectants (Frier, 1971) for a variety of clinical purposes.

Antimicrobial agents can be divided into inorganic and organic ones according to their chemical composition. Inorganic antimicrobial agents showed longer life expectancy and high heat resistance; however, they exhibit weak mould- proof activity and large dosages are needed when used (Zhou, Xia et al. 2004, Guo, Ma et al. 2005). While organic antibacterial agents showed good inhibition efficiency, a broad spectrum of activity and blending compatibility with organic matrixes such as textile, paints, polymer, etc., however, their relatively low stability cannot be ignored (Suci, Vrany et al. 1998, Sauvet, Dupond et al. 2000). In addition, the activity of all these compounds is temporary and thus requires repeated applications for a longer term biocide effect. Therefore, materials including plastics endowed with a permanent antimicrobial activity are a growing sector of the specialty biocides industry.

5.1. Polymers with antimicrobial activity

Low molecular weight antimicrobial agents suffer from many disadvantages, such as toxicity to the environment and short-term antimicrobial ability (Sauvet, Dupond et al. 2000, Jones, Djokic et al. 2005). To overcome problems associated with the low molecular weight antimicrobial agents, antimicrobial functional groups can be introduced into polymer molecules. The use of antimicrobial polymers offers promise for enhancing the efficacy of some existing antimicrobial agents and minimizing the environmental problems accompanying conventional antimicrobial agents by reducing the residual toxicity of the agents, increasing their efficiency and selectivity, and prolonging the life time of the antimicrobial agents (Samour 1976). Many studies have been performed on the antibacterial activity of low-molecular weight and polymeric quaternary ammonium salts (Kanazawa, Ikeda et al. 1993, Cakmak, Ulukanli et al. 2004). The target size of the cationic biocides is the cell envelope of bacteria; thus, an increase in the molecular size due to polymerization, which may result in reduced permeability, is not regarded as a factor seriously affecting their activity. It is worth noting that polycationic biocides have been shown to possess a higher activity against bacteria (Ikeda, Ledwith et al. 1984, Ikeda, Yamaguchi et al. 1984). In addition, polymeric biocides are particularly important because they possess promising advantages over monomeric forms (Asian and Sun 2004, Chen, Worely et al. 2004). Polycations with main chain or pendant quaternary ammonium salts show outstandingly high antibacterial activity against Gram-negative and Gram-positive strains and exhibit a wide spectrum of antimicrobial activity (Nonaka, Noda et al. 2000). Polymers substituted by quaternary ammonium salts (Gerba, Janauer et al. 1984, Avram, Lacatus et al. 2001), phosphonium salts(Kanazawa and Ikeda 2000), pyridinium cations (Li, Shen et al. 2000) and quaternized polymer (dimethlyaminoethylmethacrylate) (Lenoir, Pagnoulle et al. 2006) exhibited an antimicrobial activity. As a rule, these cationic biocides interact with the negatively charged membrane of the bacteria, which is accordingly disrupted and disintegrated.

Antimicrobial agents kill bacteria through various means depending on the types of bacteria. Most antiseptics and disinfectants kill bacteria immediately on contact by causing the bacteria cell to burst, or by depleting the bacteria's source of food preventing bacterial reproduction, also known as bacterial conjugation. Antimicrobial polymers commonly kill bacteria by this first method which is accomplished through a series of steps. First, the polymer must adsorb onto the bacteria cell wall. Most bacterial surfaces are negatively charged therefore, the adsorption of polymeric cations has proved to be more effective than adsorption of polymeric anions. The antimicrobial agent must then diffuse through the cell wall and adsorb onto cytoplasmic membrane. Small molecule antimicrobial agents excel at the diffusion step due to their low molecular weight, while adsorption is better achieved by antimicrobial polymers. The disruption of the cytoplasmic membrane and subsequent leakage of cytoplasmic constituents leads to the death of the cell. In another publication (Nonaka, Li et al. 2003), (Uemura, Moritake et al. 1999) the effect of loading biocides on polymer was studied. A longer spacer length refers to the length of carbon chain that composes the polymer backbone resulted in higher activity. There are two primary explanations for this effect. First, longer chains have more active sites available for adsorption with bacteria cell wall and cytoplasmic membrane. Second, longer chains aggregate differently from shorter chain, which again may provide a better means for adsorption. However, shorter chain length diffuses more easily.

5.2. Clay minerals with antimicrobial activity

Pharmacology studies have revealed that montmorillonite adsorbed bacteria such as *Escherichia coli (E.coli), Staphylococcus aureus (S.aureus)* and immobilized cell toxins (Schell, Lindmann et al. 1993, Herrera, Burghardt et al. 2000, Zhou, Xia et al. 2004, Hu, Xu et al. 2005). Some researchers found that natural clay minerals showed no antibacterial effect but could adsorb and kill bacteria when materials with antimicrobial activity were intercalated. There are a certain number of reports about modified MMT with antibacterial activity, such as cetylpyridinium-exchanged MMT, MMT-carrying copper and silver ions as effective bacteriostasis (Carcelli, Mazza et al. 1995, Uchida 1995, Jantova, Lauda et al. 1997, Hu and Xia 2006, Jo, Rim et al. 2007, Wang, Li et al. 2007). Modified layered silicates could adsorb both natural and anthropogenic toxin and exhibited an inhibitory property for the proliferation of bacteria (Guo, Ma et al. 2005). Antimicrobial activity of three kinds of commercially available montmorillonite nano-clays including a naturally occurring one (Cloisite Na$^+$) and two organically modified ones (Cloisite 20A and Cloisite 30B) against four representative pathogenic bacteria (two Gram-positive ones such as Staphylococcus aureus and Listeria monocytogenes, and two Gram-negative ones such as Salmonella typhimurium and E. coli O157:H7) (Hong and Rhim 2008) was found to be dependent on the type of clay mineral and microorganisms tested. Cloisite 30B showed the highest antibacterial activity followed by Cloisite 20A, however, the unmodified montmorillonite (Cloisite Na$^+$) did not show any antibacterial activity. Especially, Cloisite 30B inactivated Gram-positive bacteria completely within an hour of incubation and inactivated Gram-negative bacteria by more than 2-3 log cycles after 8 hours incubation. SEM and TEM images of cell structure indicated that the organically modified clay caused rupture of cell membrane and inactivation of

21

the bacteria. This finding of antimicrobial activity of the organo-clay would open a new opportunity to develop polymer nanocomposites with additional functionality, i.e., antimicrobial function.

5.3. Polymer-inorganic nanocomposites with antimicrobial activity

There is urgent need to develop organic-inorganic hybrid materials provided with dual antibacterial advantages of both organic and inorganic antimicrobial agents as they will become more important in the antimicrobial material market. Two types of materials have to be distinguished depending on whether the additive is temporarily trapped within the polymer (Kanazawa, Ikeda et al. 1993, Nonaka, Uemera et al. 1996, Ignatova, Labaye et al. 2003) or permanently attached to the chains (Broxton, Woodcock et al. 1983, Ikeda, Ledwith et al. 1984, Ikeda, Yamaguchi et al. 1984). Typical examples are dispersion of a low molecular mass biocide e.g. a heavy metal (Nonaka, Uemera et al. 1996) or silver (Ignatova, 2003, Kvitek, Panacell et al. 2008) within a polymer matrix, the major limitation is the possible migration and release of the antimicrobial agent. The only way of preventing this undesired effect consists in chemically bonding the active molecule to the matrix. Then, the antimicrobial action relies on the contact between the biocide and the microorganisms. The permanency of the effect depends of course on the stability of the bonding between the biocide and the polymer.

5.4. Polymer-montmorillonite nanocomposites with antimicrobial activity

There are only several reports about polymer/layered silicate nanocomposites with antimicrobial activity (Wang, 2006, Rhim, Hong et al. 2006, Wang, Dua et al. 2009, Mondal, Bhowmick et al. 2014). The polydimethyloxane/montmorillonite– chlorhexidine acetate (PDMS/ OMMT) nanocomposite films were successfully (Meng, Zhou et al. 2009) inhibited the growth of a wide variety of microorganisms, including Gram-positive bacteria, Gram-negative bacteria (*Staphylococcus aureus* (*S. aureus*) and *Escherichia coli* (*E. coli*)). The low density polyethylene (LDPE) / (clay/carvacrol) nanocomposites exhibit excellent and prolonged antimicrobial activity against *E. coli* bacteria, while LDPE/carvacrol films loss their antimicrobial functions within several days (Shemes, Goldman et al. 2015). The superior antimicrobial behavior is ascribed to the significantly higher carvacrol content. Poly(butylene adipate-*co*terephthalate) (PBAT) nanocomposites films prepared using natural

22

montmorillonite (MMT) and cetyltrimethylammonium bromide (CTAB) modified montmorillonite (CMMT) show adequate antimicrobial activity(Mondal, Bhowmick et al. 2014). A triblock copolymer based on poly(ε-caprolactone) (PCL) and 2-(N,N-diethylamino) ethyl methacrylate (DEAEMA)/2-(methyl-7-nitrobenzofurazan) amino ethyl acrylate (NBDNAcri), was anchored via cationic exchange on montmorillonite (MMT) surface. The antimicrobial activity of the nanocomposites indicates that nature of organomodifier in the clay play an important role in *B. subtilis* and *P. putida* adhesion processes. In addition, biocompatibility studies demonstrate that both PCL and PCL/MMT materials allow the culture of murine L929 fibroblasts on its surface with high viability, very low apoptosis, and without plasma membrane damage, making these materials very adequate for tissue engineering. An antimicrobial thermoplastic plasticizer based on aliphatic anhydride derivative dodecenyl succinic anhydride (DSA) for blending poly (vinyl chloride), PVC, with gelatin in presence of montmorillonite (MMT) (Haroun, Ahmed et al. 2011) exhibited high performance antimicrobial potency against Gram-positive and Gram-negative bacteria such as: *Staphylococcus aureus* (*S. aureus*), *Klebsiella pneumonia* (*K. pneumonia*), *Bacillus cereus* (*B. cereus*), *Bacillus subtilis* (*B. subtilis*) and *Escherichia coli* (*E. coli*).

5.4.1 Morphology of polymer-MMT nanocomposites vs antimicrobial activity

Recently, antimicrobial nanocomposites were prepared by intercalation polyoxyalkylene into montmorillonite at the nanometer scale followed by the reaction with different esters (Salahuddin, Badr et al. 2012 a). The effect of molecular mass of polyoxyalkylene on the activity, morphology, oxygen consumption and the flow of calcium, potassium and sodium from microbial cells of gram-negative bacteria (*Escherichia coli*), gram-positive bacteria (*B. subtillus*) and fungi (*C. albicans* and *Cryptococcus neoformans*) was studied. This nanocomposite will enable the development of low-cost antimicrobial materials with enhanced physical and engineering properties and can be applied to a wide range of domestic, health care, packaging and engineering applications in which microbial infection is a problem. The negatively charged clay surface therefore allows the polyoxyalkylene in their cation form to intercalate into the space between the clay layers. This causes layer expansion and changes the surface properties of the clay from hydrophilic to organophilic. Then the organoclay formed is reacted with esters. Therefore, it is possible to exfoliate those individual clay layers with attached antimicrobial agents into a polymer matrix to achieve uniform dispersion and to allow antimicrobial molecules to be exposed to the external surface, producing the antimicrobial nanocomposite material. After reaction of P-hydroxymethylbenzoate and methyl salicylate with D_{230}-MMT d_{001} was observed however, in D_{2000}-MMT/ P-hydroxymethylbenzoate, D_{400}- MMT/ P-hydroxymethylbenzoate and D_{230}-MMT/ 2, 4-Dihydroxymethylbenzoate the d_{001} value characteristic of polyoxyalkylene-MMT disappeared and substituted by a new weakened broad peak at lower 2θ (Fig. 8). The shift of the basal reflection to lower angle indicates the formation of intercalated nanostructure, while the peak broadening indicate the disordered intercalated or exfoliated structure.

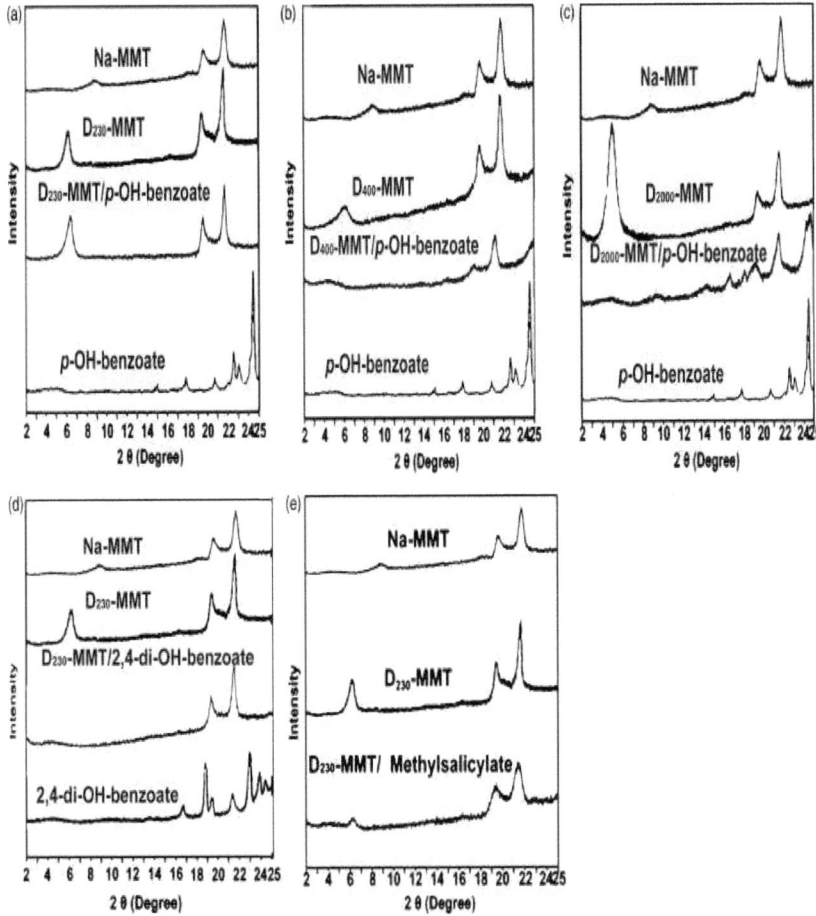

Fig.8. X-ray diffraction pattern of Na-MMT,esters, polyoxypropylene-MMT and polyoxypropylene-MMT/ esters.

The stacks of multilayers of MMT in D_{230}-MMT/ P-hydroxymethylbenzoate; the coexistence of intercalated and exfoliated morphology in D_{400}-MMT/ P-hydroxymethylbenzoate are clearly shown in Figure 9. The size of some stacks of MMT layers appears to reach about 50 nm. The stacks of MMT connect with each other edge by edge and form a flocculated structure. However, the MMT retains an exfoliated structure in D_{2000}MMT/ P-hydroxymethylbenzoate.

The ability of the nanocomposites to inhibit the growth of the selected microorganisms (*E. coli*, *B.subitilus*, *C. albicans* and *C. neoformans*) on solid media is shown in Fig.10. It was found that Na-MMT, D_{230}-MMT, D_{400}-MMT before and after reactions with esters showed no inhibition zone. However, D_{2000}-MMT/ *P*-hydroxymethylbenzoate exhibited high activity against *C. neoformans*, *C. albicans*, *E.coli* and *B.subitilus* after incubation of fungi at 30°C and bacteria at 37 °C for 24 h. It is well known that, at physiological pH, parent layered silicate exhibit a net negative charge (Nzengung, Voudrias et al. 1996). Under these conditions, negatively charged bacteria will not be significantly adsorbed onto these clays. However after modification by polyoxypropylene diamine hydrochloride with different molecular masses produce nanocomposites with different degree of hydrophobicity. Diameters of inhibition zones ranged between 10-14 mm depending on the type of microorganism. In addition, the effect on Gram positive bacterium is higher than Gram-negative bacteria. The fundamental difference between Gram positive bacterium and Gram- negative bacteria is that the latter have a so called outer membrane, which is not found in Gram positive bacteria. The different barriers differ in their ability to prevent the penetration of microbicides which explains the variation in sensitivity.

a)

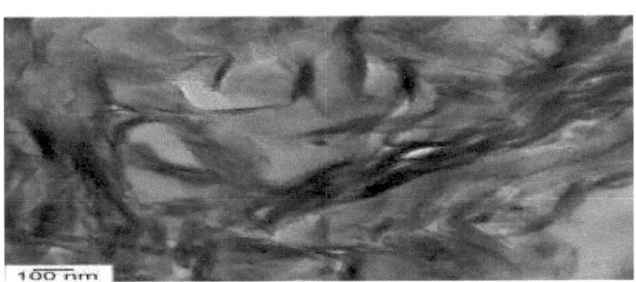

b)

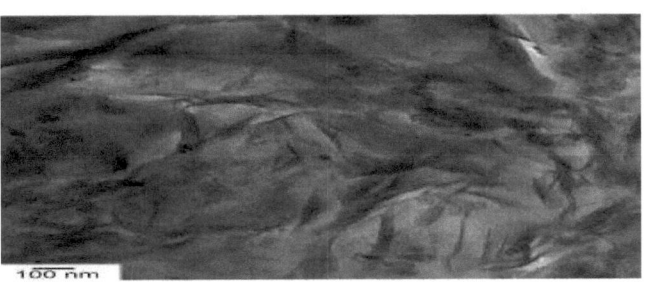

c)

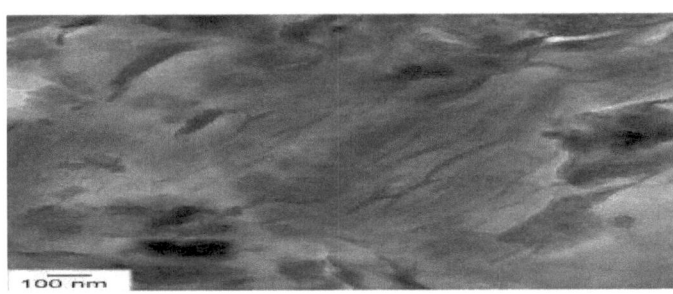

Fig.9. TEM images of a) D_{230}-MMT/p-hydroxymethylbenzoate; D_{400}-MMT/$p$hydroxymethylbenzoate and D_{2000}-MMT/p-hydroxymethylbenzoate nanocomposites.

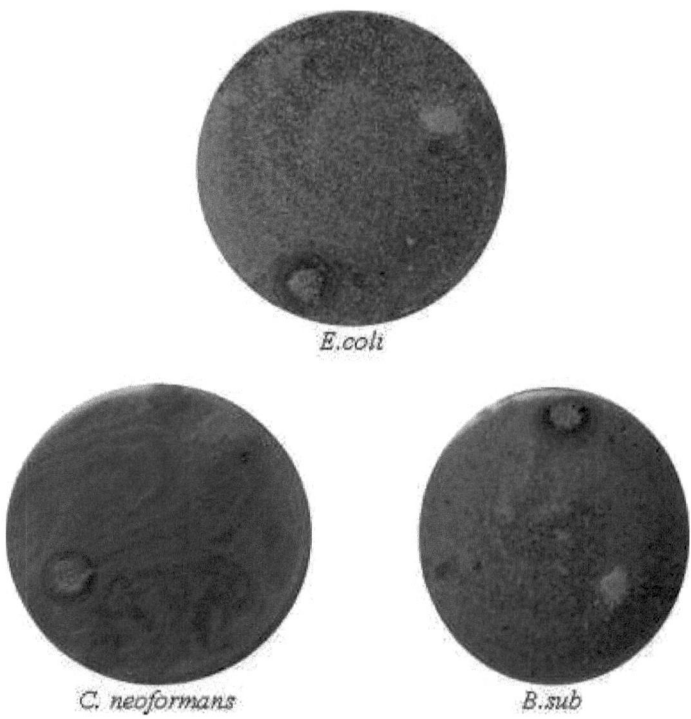

E.coli

C. neoformans   B.sub

Fig10.Inhibition zones of D_{2000}-MMT/p-hydroxymethylbenzoate nanocomposite against different species of microorganisms.

It is worth mentioning that Na-MMT/ P-hydroxymethylbenzoate show a weaker effect on all microorganisms. It seems that exfoliated nanocomposite that has higher polymer content (D_{2000}-MMT) of ca~40 wt% exhibited higher activity than intercalated nanocomposite with a lower polymer content (D_{230}-MMT, D_{400}-MMT) of ca ~ 6-18 wt%. This exfoliated nanocomposite favor the contact with microorganism, then the biocidal moiety could be transferred to or contact the microorganism in amounts sufficient to kill it.

It was reported that p-hydroxybenzoic acid alkyl esters (pKa=8.5) exhibit the antimicrobial activity at neutral and slightly higher pH values at specific concentration. As membrane active microbicides, their primary mode of action is based on the inhibition of nutrient transport into the microbial cell (Friedman,

2003). Membrane active microbicides including phenols, quaternary ammonium salts act non specifically by coating the cell wall of the microbe adsorptively (association), a process which is initially reversible by dilution, particularly when the agents are added in non lethal concentrations and redressed quickly enough (Paulus 2005). The adsorption process causes changes in the outer membrane and along the cell wall. These outer barriers eventually lose their integrity, with the result that microbicidal molecules are allowed access to the cytoplasmic membranes so that they can release their lethal effects: disarrangements in the semi-permeable properties of the cytoplasmic membrane, inhibition of enzymes localized there, escape of essential components from the cytoplasm, precipitation in the cytoplasm and finally disintegration of the cells. The growth inhibiting effect was quantitatively determined by the rate of cell death as shown in Fig. 11. The results showed that increasing the concentration from 2.5 to 10 mg/ml of D_{2000}MMT/P-hydroxymethylbenzoate increases $B.subitilus$ dead cells from 87 to 95%; $E.coli$ from 48 to 76%; $C.$ $albicans$ from 85.6 to 97.7% and $C.$ $neoformans$ from 79 to 84%. Increasing the concentration to 20 mg/ml of D_{2000}MMT/P-hydroxymethylbenzoate yielded 100% killing for all microorganisms. This leads to the suggestion that the mode of action depends on the plasma membrane of the microorganism.

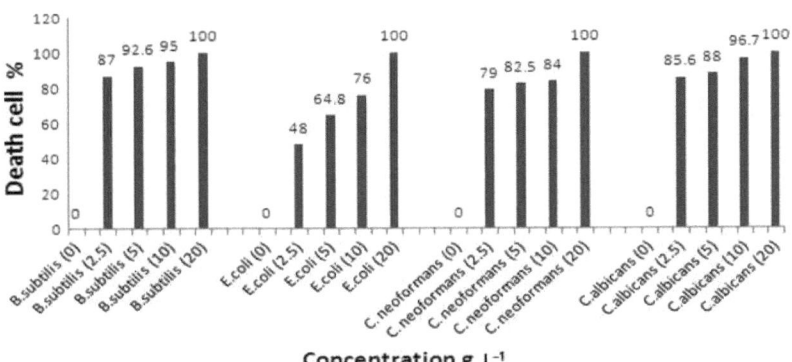

Fig.11. Percentage of dead cells of different microorganisms using different concentrations (0-20 mg/ml) of D_{2000}-MMT/ P-hydroxymethylbenzoate nanocomposite.

Fig. 12 clearly shows that *B. subtilus* was elongated after treatment with $D_{2000}MMT/P$-hydroxymethylbenzoate nanocomposite. To examine the degree of elongation, the results of quantitative analysis of the ensuing morphology are presented. The average length of untreated and treated *B. subtilus* was 21 ± 15 and 116 ± 63 μm, respectively.

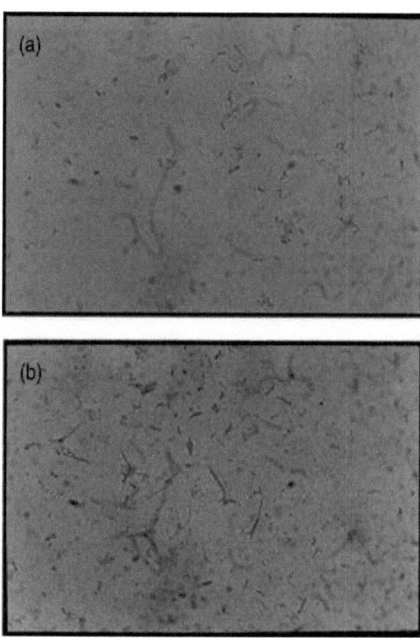

Fig.12. Morphology of: *B. Subtillus* a) before treatment; b) after treatment with D_{2000}-MMT/ *P*-hydroxymethylbenzoate nanocomposite.

Fig. 13 shows untreated and treated *C.albicans*. It appears that the treated cells were swelled and made budding. Measurements of the area of untreated cells reveal a distribution of sizes with an average of 9.5 ± 6 μm². The treated cells exhibit a wide range of area with an average of 239 ± 241μm². However, *E.coli* and *C. neoformans* show no morphological changes.

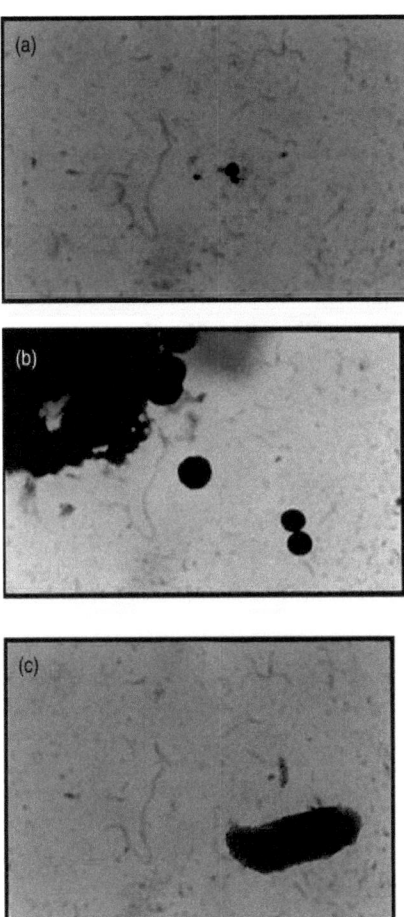

Fig. 13. Morphology of: *C.albicans* a) before treatment; b) after treatment with D_{2000}-MMT/ *P*-hydroxymethylbenzoate nanocomposite.

Oxygen consumption was measured to determine the effect of the nanocomposite on the cell metabolism mainly respiration (Salahuddin, Badr et al. 2012 a). It was observed as represented in Fig. 14 that the percentage of oxygen consumption of *B.subitilus* and *C. neoformans* is higher than the untreated cells at low concentration of the nanocomposite (2.5 mg/ml), then decrease with increasing the concentration and completely disappeared at high concentration (20 mg/ml). However, the oxygen consumption ratio in

C.albicans and *E.coli* was lower than untreated cells at low concentration of the nanocomposite (2.5 mg/ml), then increase in *E.coli* and decrease in *C.albicans* with increasing the concentration of the nanocomposite and all metabolic activities disappeared at 20 mg/ml. This indicates that the action of nanocomposite was varied according to the microorganisms. It was reported that attachment of antimicrobial substances to the surface of the cell membrane disturbing the permeability and respiration functions of the microbial cell. This disturbance leads to cell death.

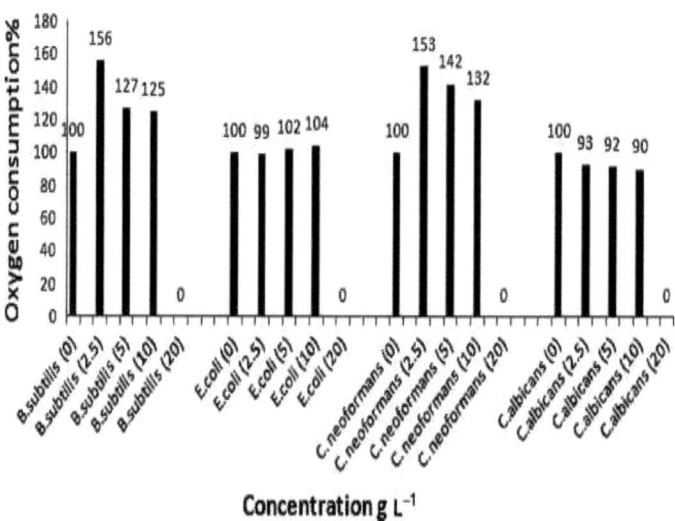

Fig.14. Effect of different concentration of D_{2000}-MMT/ *P*-hydroxymethylbenzoate nanocomposite on the oxygen consumption of different microorganisms.

The results in Fig. 15 cleary showed the effect of nanocomposite on Ca^{++}ions influx from the various tested microorganisms. It was observed that, D_{2000}-MMT/*P*-hydroxy methylbenzoate nanocomposite increase the Ca^{++}ions influx outside the cell of *B.subitillus* however; Ca^{++}ions influx inside the cell was decreased at 2.5 mg/ml concentration of the nanocomposite then increased at higher concentration. The Ca^{++}ions influx outside the cell of *E.coli* was

32

decreased then increased by increasing the nanocomposite concentration and Ca^{++}ions influx decrease inside the cell at all concentrations. In *C.albicans*, Ca^{++}ions influx inside the cell was decreased with increasing concentration. However, Ca^{++}ions influx outside the cell was increased then decreased with increasing the nanocomposite concentration. In *C. neoformans*, the Ca^{++}ions influx was equal inside and outside the cell at 20 mg/ml of nanocomposite. The influx of Ca^{++}ions from the microorganisms leads to the death of the cell (Kvitek, Panacell et al. 2008).

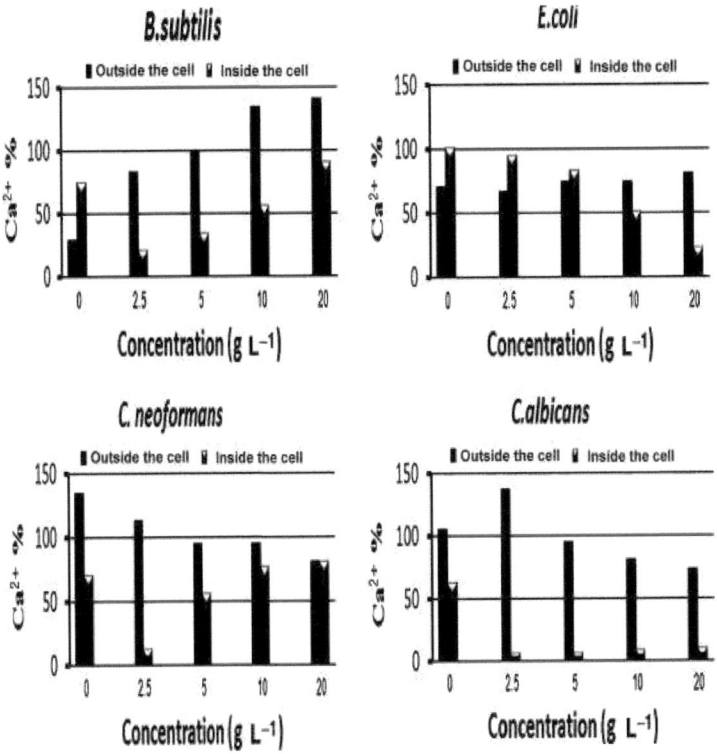

Fig. 15. Effect of different concentration of D$_{2000}$-MMT/*P*-hydroxymethylbenzoate nanocomposite on the amount of Ca^{++} inside and outside the cells in different microorganisms.

Fig.16 showed that, in treatment with D_{2000}-MMT/p-OH-benzoate the concentration of K^+ ions increase outside the cell then decrease with increasing the concentration of the nanocomposite in all microorganisms.

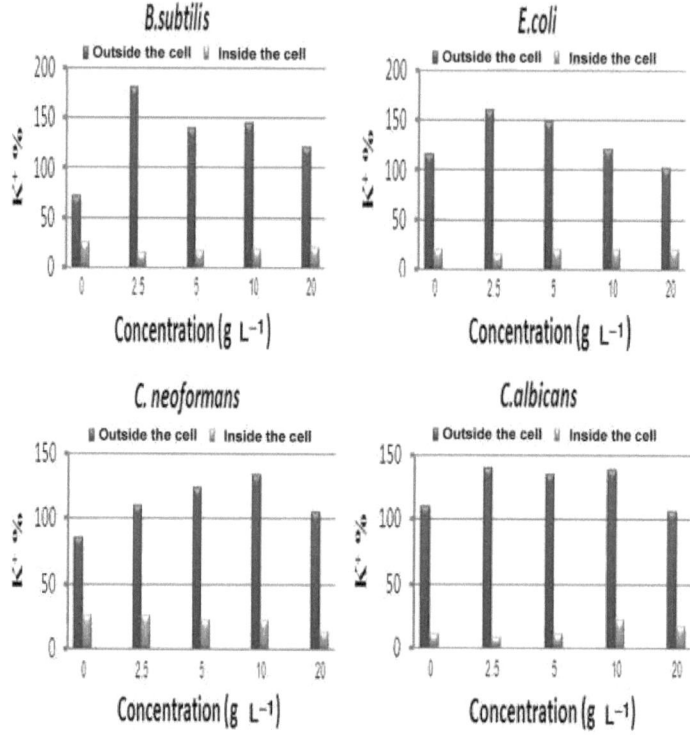

Fig.16. Effect of different concentration of D_{2000}-MMT/P-hydroxymethylbenzoate nanocomposite on the amount of K^+ inside and outside the cells in different microorganisms.

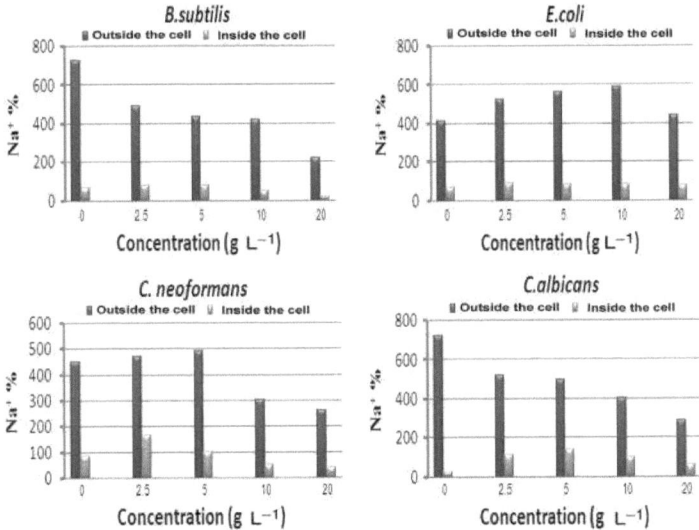

Fig.17. Effect of different concentration of D_{2000}-MMT/*P*-hydroxymethylbenzoate nanocomposite on the amount of Na^+ inside and outside the cells in different microorganisms.

The results in Fig.17 indicated that, the D_{2000}-MMT/*P*-hydroxymethylbenzoate nanocomposite caused flow of sodium ions (Na^+) ions from the microbial cells. The flow of (Na^+) ions was affected by the concentration of nanocomposite and the sensitivity of tested organism. The values of the flow of sodium ions (Na^+) ions decreased outside the cell of *B.Subtillus* and *C.albicans* and increase then decrease inside the cell with increasing the concentration of the nanocomposite. In *E.coli* and *C.neoformans*, concentration of Na^+ions increase then decrease outside and inside the cell. It is well known that Ca^{+2}, K^+ and Na^+ ions control different vital process of the cell e,g, cell replication, co-factor in many enzymes and signal transduction.

It was reported that surface-active microbicides with cationic properties include aliphatic diamines and quaternary ammonium compounds are attracted by the negatively charged surface of the microbial cell and particularly strongly adsorbed by ionic interaction with phospholipids of the cell wall. The cell wall thereby losses its function as a protective barrier so that the active ingredients are able to penetrate to the cytoplasmic membrane. The cationic microbicides impair permeability until the cytoplasmic membrane no longer functions as a semi-permeable membrane (Denyer, 1995). It was generally accepted that the mode of action of the antimicrobial nanocomposite is as follows: The bacterial membranes are negatively charged and nanocomposite positively charged. Electrostatic interactions between the cell membrane and nanocomposite enable the latter to bind to the microbial cell and facilitate the attachment of phenolic ester biocide to bacteria leading to loss of constituents of the cell and death of the cell.

In another publication (salahuddin, Badr et al. 2012b) antimicrobial schiff base nanocomposites were prepared by the reaction of ($D_{230-2000}$–MMT) nanocomposite with aldehydes. It was noted that D_{2000}-MMT / vanillin Schiff base nanocomposites shows some highly delaminated structures. However, D_{230}-MMT / vanillin and D_{400}MMT / vanillin showed the coexistence of both intercalated (stacks of multilayers of MMT) and exfoliated structure. Na-MMT, D_{230}-MMT/vanillin, D_{400}MMT/vanillin schiff base cannot inhibit the growth of microbes however; a mixture of Na-MMT/ aldehydes showed small inhibitory effect and D_{2000}MMT/vanillin Schiff base showed a higher inhibitory effect. The inhibitory actions are observed against a wide variety of microorganisms, including Grampositive bacteria, Gram-negative bacteria and Fungi. Diameters of inhibition zones were ranged between 13-22 mm on Fungi, 15 mm on Gram-negative bacteria and 20 mm on Gram-positive bacteria growth after incubation for 48 h. The same result was observed in chitosan-organic clay nanocomposites that showed the good inhibitory for Gram positive bacteria growth, but little effect on Gram –negative bacteria (Rhim, Hong et al. 2006). In Gram- positive bacteria, there are networks with plenty of pores which allow foreign molecules to come into the cell without difficulty (Casterton and Cheng 1975). On the other hand, in Gram-negative bacteria the outer membrane (peptidoglycan layer) is a potential barrier against foreign molecules with high molecular weight. The changes in peptedology can lead to change in microbial forms.

The inhibition percentage of D_{2000}-MMT/vanillin Schiff base was assayed with different concentrations (0, 2.5, 5, 10 and 20 mg/ml). The 20 mg/ml concentration performed zero surviving ratios on cells. At sub MICs dose (10 mg/ml) the effect of D_{2000}-MMT/vanillin Schiff base on Gram positive bacteria more than Gram negative bacteria and the survival percentage of *B.subtillus* and

E.coli were 4 % and 12 %, respectively. For fungi, the survival percentage of *C. neoformans* and *C.albicans* were 36 % and 15 %, respectively. It was reported that D_{2000}-MMT/vanillin Schiff base affected on respiration of *C .albicans* and *E. coli*. It increases the respiration percentage however; it decreases the respiration percentage of *B. subitilus* and *C. neoformans* than untreated cells. The nanocomposite affect on electron transport system and cytochromes which controlled the oxygen consumption. It is worth noting that the highest consumption of oxygen in *E. coli* and *C. albicans* was found at 2.5 mg/ml concentration due to resistance of cells for death. The amount of Ca^{+2}, K^+ and Na^+ ions outside and inside the cells was disturbed. This indicates that D_{2000}MMT/vanillin Schiff base affect on cytoplasmic membrane leading to changing the equilibrium by increasing or decreasing the amount of ions outside and inside the cells. The reaction with cytoplasmic membrane leading to formation of pores selective for K^+ ions, Na^+ ions and inorganic phosphate. The loss of those ions resulted in immediate dissipation of the membrane potential leading to cell death (Avram, Lacatus et al. 2001).

It is worth mention that TEM of D_{2000}-MMT/vanillin Schiff base shows smaller nanodomains than D_{230}-MMT/vanillin, D_{400}-MMT/vanillin, therefore larger surface area will be obtained and more bacteria are adsorbed and immobilized on the surface of this nanocomposite. In this way, D_{2000}MMT/vanillin Schiff base can exert strong inhibitory effect on growth of microorganisms. The presence of O-H and $–OCH_3$ groups in vanillin promotes the interaction between the cell and the nanocomposite.

6.Antimicrobial, toxicity, and anti-inflammatory properties in experimental animals using polyamide-montmorillonite nanocomposites

There are various problems arising from the use of antimicrobial compounds such as local tissue irritation, interference with wound healing process, hypersensitivity reactions, systemic toxicity, narrow antimicrobial spectrum and emergence of resistance. Recently a series of polyamide containing chemically bound 1,3,4-oxa (thia) diazoles have been prepared (Salahuddin, El-Barbary et al. 2014) antimicrobial, toxicity and anti-inflamatory properties in experimental animals were studied. The ^{1}H-NMR spectral data was found to be in quite agreement with the proposed structure (Scheme 1). Figure 18 shows the XRD pattern of nanocomposites. The characteristic peaks of 001 plane for X-MMT, XI-MMT, and XII-MMT are observed at 5.84, 5.8, 6.14° with basal spacing of 15.15, 15.23, and 14.38 A° respectively.

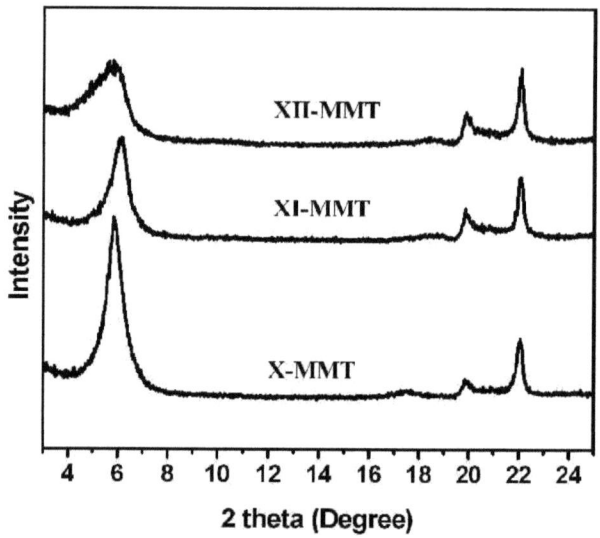

Fig. 18. X-ray diffraction pattern of polyamide-MMT nanocomposites.

According to the Bragg's law, the peak shifting from lower d-spacing (9.6 A°) characteristic to Na-MMT to higher one is due to the intercalation of polymer into the interlayer of MMT. More direct evidence for the formation of a nanocomposite is provided by the TEM of film produced from cast suspension of nanocomposite in acetone. TEM images of X-MMT, XI-MMT, and XIIMMT (Fig.19 (a–c)) display individual silicate layers appear as dark lines. There are some irregular dispersions of the silicate layer in X-MMT and XII-MMT. Some particles maintained their original ordering while others were exfoliated. This is consistent with the observation of XRD studies shown in Fig.18.

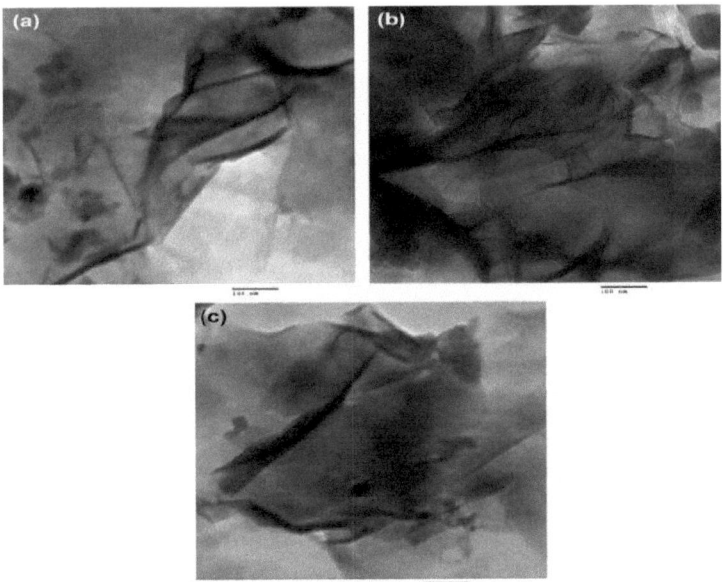

Fig. 19. TEM of a) X-MMT, b) XI-MMT, c) XII-MMT.

However, TEM image of XI-MMT indicates that the separation between the dispersed plates is higher than in X-MMT and XII-MMT nanocomposites. This TEM photograph proves that the most clay layers were exfoliated and dispersed homogeneously into polyamide matrix. Considering the proceeding results, the existing morphology could be affecting the release behavior of heterocyclic compounds.

The controlled release tests were carried out by suspending the polyamide (VII, VIII, and IX) and their nanocomposites (X-MMT, XI-MMT, and XIIMMT) in the buffer solution at pH 2.3, 5.8, and 7.4. The amount of heterocyclic compounds (I, II and III) released from polyamide (VII, VIII, and IX) was measured at 2880 min of time intervals by measuring the absorbance at λ_{max} = 343 nm, λ_{max} = 301 nm and λ_{max} = 248 nm, respectively. The results (Figure 20) showed that the release behavior depends on the pH of the medium in the order 7.4 > 5.8 > 2.3; 50% of I was released after 10 hr, 15hr, and 19 hr at pH 7.4, 5.8, and 2.3, respectively. The release of II reaches 90% after 30 min, 1.5 hr, 2 hr at pH 7.4, 5.8, and 2.3. The release of III reaches 90% at pHs 5.8 and 7.4 after 24 hr and 40% after 25 hr at pH 2.3. In the slightly basic medium, the hydrolysis of the amide linkages takes place through the attack of the nucleophile on the electron-deficient carbonyl.

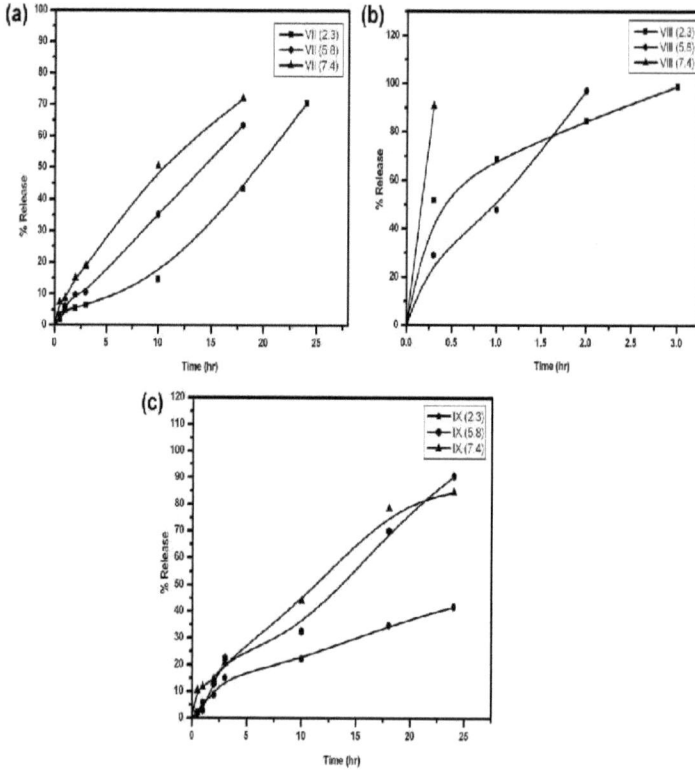

Fig.20. In vitro release 1,3,4-oxa (thia)-diazoles (I, II, III) from polyamides (VII, VIII, IX) polyamides at different pH (2.3, 5.8, 7.4).

Fig. 21 shows the release of I, II, and III from X-MMT, XI-MMT, and XII-MMT at different pH. Comparing release of 1,3,4-oxa(-thia)diazoles (I, II, III) from polyamides and intercalated nanocomposites of the same polymer presented in Figure 20 shows that linear polymers were characterized by faster rates than the intercalated polymers. The increase in the rates was attributed to higher interactions of polymers with the used medium than the intercalated polymers. Also the higher content of 1,3,4-oxa(-thia)diazoles in linear polyamides leads to increase the rate of release. This slow release process may be interpreted based on the good barrier properties (Salahuddin, Abo-El-Enein et al. 2010) due to the tortuous diffusion pathways that small molecules must travel in order to clear the material. This explained that nanocomposite formation resulted in sustained release of heterocyclic compounds in comparison with the polymer. Fig. 21 revealed that 10%, 32% and 16% of I, II, and III was released after 24 hr, respectively at pH 2.3.

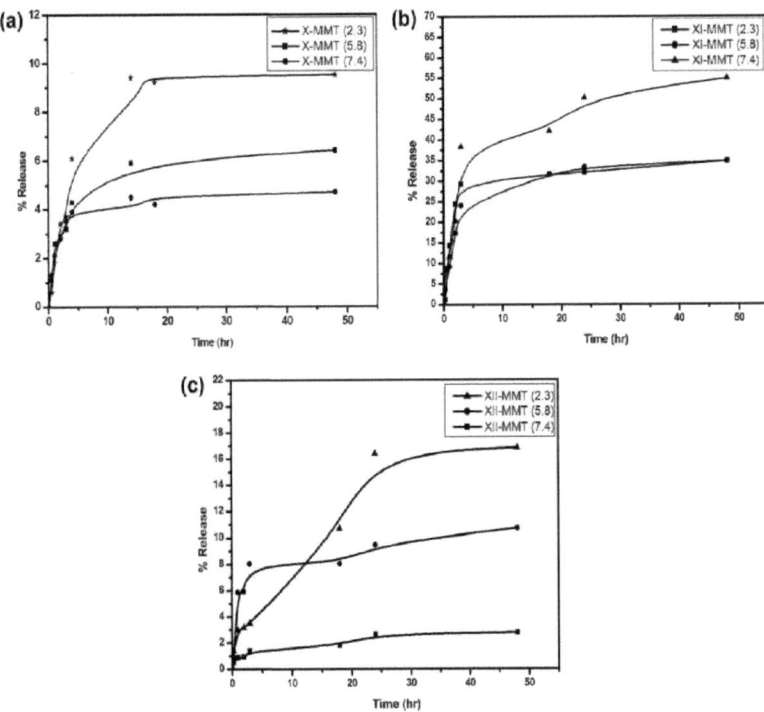

Fig.21. In vitro release of 1,3,4-oxa (thia)-diazoles (I, II, III) from a) X-MMT; b) XI-MMT; c) XII-MMT polyamide-MMT nanocomposites at different pH(2.3, 5.8, 7.4).

A comparison between the release rates from X-MMT, XI-MMT, XII-MMT suggest that the rate increases with increasing the hydrophilic characters of polyamides loaded 1,3,4-(oxa)diazoles. Presence of NH-CO-NH in XI-MMT increases the surface area in contact with the hydrolyzing medium than X-MMT, XII-MMT that contains S-CO-NH group. All the synthesized compounds were screened for their antimicrobial activity. The obtained results illustrated in Figure 22 and Table I showed that I, II, III, VII, VIII, and IX had a significant antimicrobial activity against tested microorganisms with variable inhibition zones expressed as mm. The highest antimicrobial activity of compound III than I and II may be related to the presence of two sulfur atoms in the molecule, one inside the hetero ring and the other as amidothio besides the presence of electron withdrawing (chlorine) atom. It is interesting to note that a minor change in the molecular structure of investigated compounds may have a pronounced effect on antimicrobial screening (Shetty, Khazi et al.2010). It is worth mention that polyamides loaded 1,3,4-(oxa) diazoles (VII, VIII, IX) were more active against tested microorganisms than 1,3,4-(oxa) diazoles (I, II, III). This may be explained by the presence of NH_2 and amide groups that affect the tested microorganisms. However, the results clearly revealed that the intercalation of polymers into clay (X-MMT, XI-MMT, XII-MMT) exhibit no *in vitro* antimicrobial activity against all tested microorganism. This may be attributed to the good barrier of clay mineral (Salahuddin, Abo-El-Enein et al. 2010) that decreases the contact between the functional groups and the tested microorganisms in this media. The most potent antimicrobial compounds were investigated to detect their MIC. Different concentrations of the most active antimicrobial compounds (VII and IX) were screened to detect their MICs and MBCs values against A. hydrophila (the most sensitive one). The obtained results showed that the MIC and MBC values of VII and IX were recorded at 12.5 and 25 mg/mL (MICs) and were 6.25 and 3.16 mg/mL (MBCs), respectively. However, tetracycline which used as a standard antimicrobial drug needed lower concentration (0.012 and 0.0062 mg/mL) to give the same effect.

Table 1. Inhibition zones (mm) of tested compounds against different microorganisms

Tested Compounds	Inhibition zone (mm) / Tested Microorganisms						
	B. subtilis	Shegilla sp	A. hydrophila	S. aureus	E.coli	C.albicans	A.niger
I	13.4±0.03	11.5±0.05	17.2±0.23	16.2±0.03	14.3±0.08	14.7±0.06	13.7±0.03
VII	16.7±0.03	22.7±0.18	19.8±0.03	22.00±0.05	18.20±0.23	17.80±0.03	28.00±0.05
VII-MMT	0.00	0.00	0.00	0.00	0.00	0.00	0.00
II	0.00	18.30±0.08	0.00	11.30±0.06	12.30±0.10	0.00	10.60±0.06
VIII	24.30±0.33	19.20±0.08	19.50±0.05	20.80±0.14	15.70±0.10	9.70±0.12	19.50±0.05
VIII-MMT	0.00	0.00	0.00	0.00	0.00	0.00	0.00
III	20.80±0.03	15.70±0.19	26.30±0.19	19.20±0.08	19.70±0.10	19.70±0.06	16.50±0.05
IX	23.80±0.10	23.00±0.00	26.20±0.10	20.5±0.05	25.20±0.03	24.20±0.10	38.30±0.03
IX-MMT	0.00	0.00	0.00	0.00	0.00	0.00	0.00
ANOVA F	260.86	322.05	377.00	711.81	270.61	826.64	3939.38
ANOVA P value	0.00	0.00	0.00	0.00	0.00	0.00	0.00

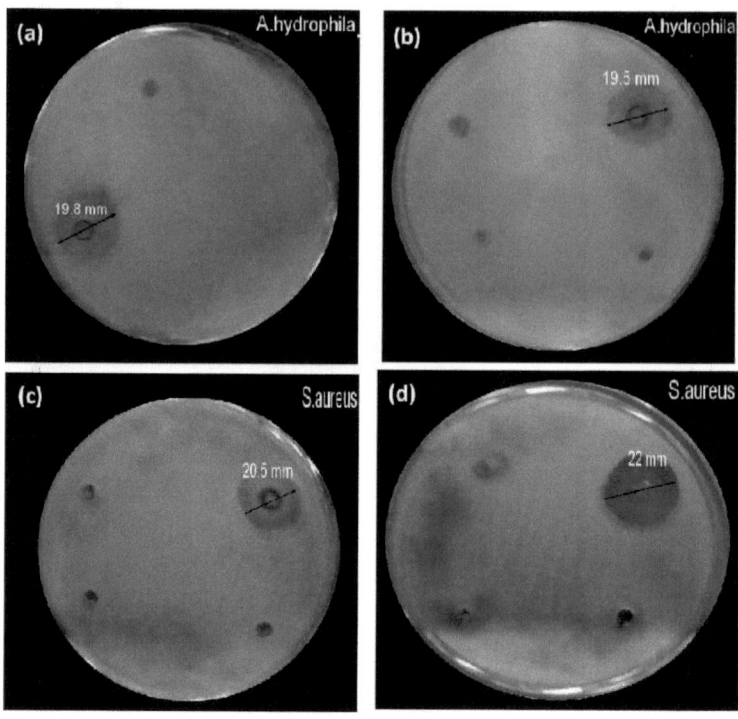

Fig. 22. Inhibition zones of (a) VII for A. hydrophila (b) VIII for A. hydrophila (c) IX for S. aureus, (d) VII for S. aureus.

Statistical analysis (Table I) showed that the effect of the tested compounds on the tested micro-organisms were highly significant at probability value (P < 0.01) on diameters of inhibition zones. The preliminary results presented in *in vitro* study suggested that VII, IX may offer good antimicrobial activity.

Infection was induced by topical application of the bacteria into burn site on the animal's backs and treated with the same manner. The ability of tested compounds (VII, IX, X-MMT and XII-MMT) to inhibit the spread of bacteria from infected burn wound to other organs were compared to untreated control and Mebo treated groups. The results in Table II show that the number of

44

bacteria in blood increased after two days than in organs due to its transport from skin to blood. All tested compounds revealed significant antimicrobial activities as they reduced S. aureus load in blood comparing to Mebo and control treatments. The number of bacteria was decreased with increasing the time post infection. This indicated the ability of investigated treatment to prevent or reduce sepsis.

Table 2. Number of bacterial load in blood and organs (skin, Spleen and liver) after different days.

Code	skin			Blood			Spleen			liver		
	2 days	4 days	6 days	2 days	4 days	6 days	2 days	4 days	6 days	2 days	4 days	6 days
Control	181	193	210	214.50± 2.12	333.00± 4.75	449.50± 23.33	169.50 ± 4.95	258.00 ± 11.31	316.50 ± 23.33	200.00 ± 141.42	358.00 ± 35.36	449.50 ± 23.33
Na-MMT	150	172	189	132.50± 0.71	166.00± 0.00	216.50± 23.33	165.00 ± 2.83	179.50 ± 4.95	231.50 ± 2.12	134.00 ± 1.41	175.00 ± 35.36	299.50 ± 47.38
Mebo	135	165	200	183.00± 24.04	233.00 ± 0.00	350.00± 0.00	167.00 ± 1.41	185.00 ± 11.31	249.50 ± 23.33	136.50 ± 23.33	166.50 ± 47.38	383.00 ± 24.04
VII	100	50	30	169.50± 4.95	47.00 ± 4.24	0.00 ± 0.00	166.50 ± 23.33	90.00 ± 56.57	30.00 ± 4.24	150.00 ± 212.13	66.00 ± 0.00	000.00 ± 00.00
IX	95	45	32	170.50± 6.36	28.00 ± 7.07	0.00± 0.00	158.00 ± 59.40	144.50 ± 40.31	66.00 ± 0.00	146.50 ± 4.95	38.50 ± 7.78	000.00 ± 00.00
VII-MMT	60	33	15	116.50± 4.95	13.00 ± 4.24	0.00 ± 0.0	151.50 ± 2.12	66.00 ± 0.00	21.50 ± 12.02	183.00 ± 70.71	75.00 ± 35.36	049.50 ± 23.33
IX-MMT	70	40	12	169.50± 19.09	45.00 ± 7.07	0.00 ± 7.07	101.50 ± 2.12	66.50 ± 23.33	28.00 ± 2.83	199.50 ± 23.33	58.00 ± 11.31	033.00 ± 00.00

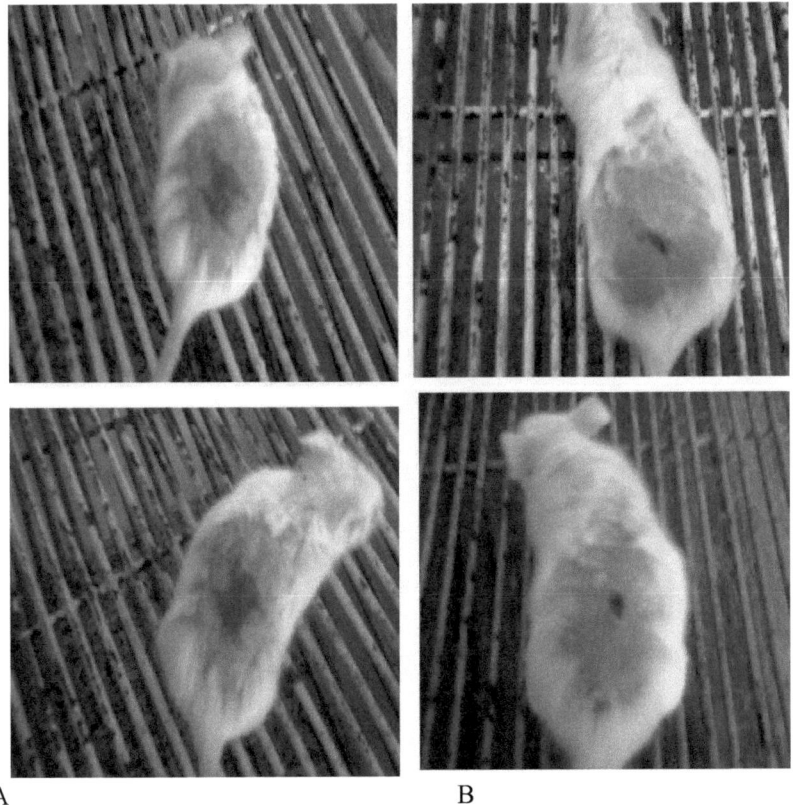

A   B

Fig.23. Mice treatment by VII (Top), IX (Bottom) of burnt skin after A) 2 days and B) 6 days.

The results revealed that compounds X-MMT and XII-MMT had the highest reduction level of S. aureus on skin, blood and spleen after 6 days of infection compared to that of untreated control group. The data reflect that these nanocomposites were the most effective treatment as it reduced bacterial load in spleen than other treatments. VII and IX possessed good antibacterial activity as they reduce bacterial load in liver when compared to other treatments.

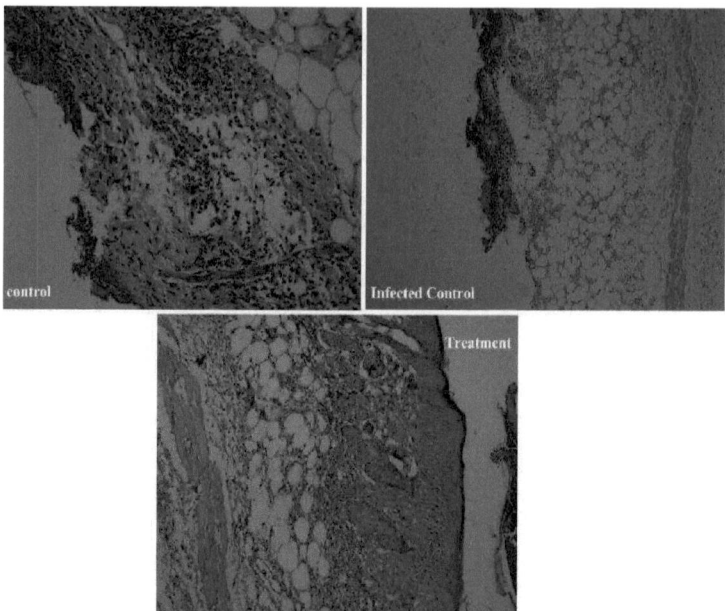

Fig. 24. Photograph of skin specimen from mice of control (burnt skin), infected control (burnt skin infected with S.aureus) and after treatment with VII.

Toxicity of compounds on shaved mouse healthy skin was reported (Hashim 2014). The mice in all treated and untreated groups did not show any sign of irritation (rash, inflammation, swelling, scaling, and abnormal tissue growth in the affected area) when observed for a period of seven days (Fig. 23). Control stained with hematoxylin and cosin (Fig. 24) showed ulcer in the epidermis and dermis layers with inflammatory infiltrate in the dermis layer. However, treated skin showed normal layers structure. In addition, wounds treated with X-MMT and XII-MMT healed faster than untreated wounds. The results presented in this thesis suggested that VII, IX, X-MMT, and XII-MMT nanocomposites offer a promising and novel means for the treatment of *S. aureus* burn wound infection and open new insight into further investigation on this area.

References

Aguzzi, C., P. Cerezo, C. Viseras and C. Caramella (2007). "Uses of clays as drug delivery systems: possibilities and limitations." Appl Clay Sci **36**(1): 22-36.

Akelah, A. and A. Moet (1994). "Polymer–clay nanocomposites: free-radical grafting of polystyrene on to organophilic montmorillonite interlayers". J Mater Sci **31**: 3589–3596.

Akelah, A., N. Salahuddin, A. Hiltner, E. Baer and A. Moet (1994). "morphological hierarchy of butadieneacrylonitrile/montmorillonite nanocomposite." Nanostructured Materials **4**(8): 965-978.

Ambrogi, V. (2001). "Microporous material from kanemite for drug inclusion and release." Farmaco **56**(5-7): 421-425.

Asian, L. and G. Sun (2004). "Durable and regenerable antimicrobial textiles: Improving efficacy and durability of biocidal functions." J Appl Polym Sci **91**(4): 2588-2593.

Avram, E., C. Lacatus and G. Mocanu (2001). "Polymers with pendent functional groups: VII. Polysaccharide derivatives containing viologen groups." Eur Polym J **37**(9): 1901-1909.

Biase, S. D. (1980). "Antimicrobial activity of chlorhexidine-containing compounds." Riv Ital Stomatol **49**(9): 597-607.

Bloomfield, S. F. (1996). Hand book of disinfectants and antiseptics. Chlorine and iodine formulations. J. M. Ascenzi. New York, Marcel Dekker, Inc. 133.

Broxton, P., P. M. Woodcock and P. Gilbert (1983). "A study of the antibacterial activity of some polyhexamethylene biguanides towards Escherichia coli ATCC 8739" J Appl Bacteriol. **54**(3): 345-353.

Cakmak, I., Z. Ulukanli, M. Tuzeu, S. Karabuga and K. Genctav (2004). "Synthesis and characterization of novel antimicrobial cationic polyelectrolytes." Eur Polym J **40**: 2373-2379.

Carcelli, M., P. Mazza, C. Pelizzi and G. Pelizzi (1995). "Antimicrobial and genotoxic activity of 2, 6-diacetylpyridinebis (acylhydrazones) and their complexes with some first transition series metal ions, X-ray crystal structure of a dinuclear copper (II) complex." J Inorg Biochem **57**: 43-62.

Casterton, J. W. and K. J. Cheng (1975). "The role of the bacterial cell envelope in antibiotic resistance." J Antimicrob Chemother **1**: 363-377.

Chen, Y., S. D. Worely, T. S. Huang, J. Weese, J. Kim, C. I. Wei and J. F. Williams (2004). "Biocidal polystyrene beads. IV. Functionalized methylated polystyrene." J Appl Polym Sci **92**(1): 368-372.

Cypes, S. H. W. and M. Saltzman (2003). "Organosilicate polymer drug delivery systems: controlled release and enhanced mechanical properties." J Control Release **90**: 163-169.

Denyer, S. P. (1995). "Mechanisms of action of antibacterial biocides" International biodeterioration & biodegradation **36** (3-4): 227-245.

Dong, Y. and S. S. Feng (2005). "Poly (D, L-lactide-co-lycolide)/montmorillonite nanoparticlesfor oral delivery of anticancer drugs." Biomaterials **26**: 6068-6076.

Fejer, I., M. Kata, I. Eros, O. Berkesi and I. Dekani (2001). "Interaction of monovalent cationic drugs with montmorillonite." Colloid Polym Sci **280**: 372-379.

Feng, S. S., L. Mei, P. Anitha, C. W. Gan and W. Zhou (2009). "Poly(lactide)–vitamin E derivative/montmorillonite nanoparticle formulations for the oral delivery of Docetaxel." Biomaterials **30**: 3297-3306.

Friedman, M., P. R. Henika and R. E. Mandrell (2003). "Antibacterial Activities of Phenolic Benzaldehydes and Benzoic Acids against Campylobacter jejuni, Escherichia coli, Listeria monocytogenes and Salmonella enteric." J food protection **10**: 1752-1948.

Frier, M. (1971). Inhibition and destruction of the microbial cell. Derivatives of 4-amino-qunadinium and 8-hydoxyquinoline. W. B. H. eds. London, England, Academic Press Ltd. 107.

Gerba, C. P., G. E. Janauer and M. Costello (1984). "Removal of poliovirus and rotavirus from tap- water by a quaternary ammonium resin." Water Res **18**(1): 17-19.

Giordano, C., V. Sanginario, L. Ambrosio, L. D. Silvio and M. Santin (2006). "Chemical–physical characterization and in vitro preliminary biological assessment of hyaluronic acid benzyl esterhydroyapatite composite." J Biomater App **20**: 237-252.

Guo, T., Y. L. Ma, P. Guo and Z. R. Xu (2005). "Antibacterial effects of the Cu(II)-exchanged montmorillonite on Escerichia coli K88 and Salmonella Choleraesuis." Vet microbial **105**(2): 113-118.

Haraguchi, K., T. Takehisa and M. Ebato (2006). "Control of cell cultivation and cell sheet detachment on the surface of polymer/clay nanocomposite hydrogels." Macromolecules **7**: 3267-3275.

Haroun, A. A., E. F. Ahmed and M. A. A. El-Ghaffar (2011). "Preparation and antimicrobial activity of poly (vinyl chloride)/gelatin/montmorillonite biocomposite films." The Journal of Materials Science: Materials in Medicine **22**(11): 2545-2553.

Hashim, A. (2014) Loading of some 1,3,4-Oxa(thia)diazole derivatives on polymers/clay nanocomposites and their applications M.Sc. Thesis, Tanta University, Faculty of Science, Chemistry Department..

Herrera, P., R. C. Burghardt and T. D. Phillips (2000). "Adsorption of salmonella enteritidis by cetylpyridinium-exchanged montmorillonite clays." Vet Micobiol **74**: 259-272.

Hong, S. I. and J. W. Rhim (2008). "Antimicrobial activity of organically modified nano-clays." J Nanosci Nanotechnol **8**(11): 5818-5824.

Hu, C. H., M. S. Xia (2006). "Adsorption and antibacterial effect of copper-exchanged montmorillonite on Escherichia coli K88." Appl Clay Sci **31**: 180-184.

Hu, C. H., Z. R. Xu and M. S. Xia (2005). "Antibacterial effect of Cu^{+2}exchanged montmorillonite on Aermonas hydrophila and discussion on its mechanism." Vet Microbiol **109**: 83-88.

Ignatova, M., D. E. Labaye, S.Lenoir, D.Strivay, R. Jerome and C.Jerome (2003). "Immobilization of silver in polypyrrole/ polyanion composite coatings: preparation, characterization, and antibacterial activity." Langmuir **19** (21): 8971–8979.

Ikeda, T., A. Ledwith, C. H. Bamford and R. A. Hann (1984). "Interaction of a polymeric biguanide biocide with phospholipid membranes." Biochem Biophys Acta **769**(1): 57-66.

Ikeda, T., H. Yamaguchi and S. Tazuke (1984). "New polymeric biocides: synthesis and antibacterial activities of polycations with pendant biguanide groups antimicrobial agents." Chemother **26**(2): 139-144.

Jantova, S., J. Lauda, V. Vollek and M. Zastkova (1997). "Antimicrobial effects of the macrocyclic Cu (II) tetraanhydroanino benzaldehyde complex." Folia Microbiol Prague **42** (4): 324-326.

Jo, S. C., A. R. Rim, H. J. Park, H. G. Yuk and S. C. Lee (2007). "Combined treatment with silver ions and organic acid enhances growth-inhibition of Escherichia coli 0157:H7." Food control **18** (10) 1235-1240.

Jones, D. S., J. Djokic and S. P. Gorman (2005). "The resistance of polyvinyl pyrrolidone–Iodine–poly(caprolactone) blends to adherence of Escherichia coli." Biomaterials **26**(14): 2013-2020.

Joshi, G. V., B. D. Kevadiya and H. C. Bajaj (2010). "Design and evaluation of controlled drug delivery system of buspirone using inorganic layered clay mineral." Microporous and Mesoporous Materials **132**: 526-530.

Joshi, G. V., B. D. Kevadiya, H. A. Patel, H. C. Bajaj and R. V. Jasra (2009). "Montmorillonite as a drug delivery system: intercalation and in vitro release of Timolol maleate." International Journal of Pharmaceutics **374**: 53-57.

Joshi, G. V., H. A. Patel, H. C. Bajaj and R. V. Jasra (2009 a). "Intercalation and controlled release of vitamin B6 from montmorillonite–vitamin B6 hybrid." Colloid and Polymer Science **287**(9): 1071-1076.

Joshi, G. V., H. A. Patel, B. D. Kevadiya and H. C. Bajaj (2009 b). "Montmorillonite intercalated with vitamin B as drug carrier." Applied Clay Science **45**(4): 248-253.

Kanazawa, A. and T. Ikeda (2000). "Chem Inform abstract: multifunctional tetracoordinate phosphorus species with high self-organizing ability." Coord Chem Rev **198**: 117-131.

Kanazawa, A., T. Ikeda and T. Endo (1993). "Novel polycationic biocides: Synthesis and antibacterial activity of polymeric phosphonium salts." J Polym Sci Part A: Polym Chem **31**(2): 335-343.

Kevadiya, B. D., G. V. Joshi and H. C. Bajaj (2010). "Layered bionanocomposites as carrier for procainamide." Int J Pharm **388**(1-2): 280-286.

Kvitek, L., A. Panacell, J. Soukupova, M. Kolar, R. Vecaova, R. Prucek M.

Holecová and R. Zbořil(2008). "Effect of Surfactants and Polymers on Stability and Antibacterial Activity of Silver Nanoparticles (NPs)." Phy chem **112**(15): 5825–5834.

Langer, R. S., E. U. Kathryn, S. M. Cannizzaro and K. M. Shakesheff (1999). "Polymeric systems for controlled drug release." Chemical Reviews **99**: 3181-3198.

Lee, W. F. and Y. T. Fu (2003). "Effect of montmorillonite on the swelling behavior and drug-release behavior of nanocomposite hydrogels." J Appl Polym Sci **89**(13): 3652-3660.

Lee, W. F. and L. L. Jou (2004). "Effect of the intercalation agent content of montmorillonite on the swelling behavior and drug release behavior of nanocomposite hydrogels." J Appl Polym Sci **94**(1): 74-82.

Lee, Y. H., T. F. Kuo, B. Y. Chen, Y. K. Feng, Y. R. Wen, W. C. Lin and F. H. Lin (2005). "Toxicity assessment of montmorillonite as a drug carrier for pharmaceutical applications: yeast and rats model." Biomedical Engineering Applications Basis Communications **17**: 72-78.

Lenoir, S., C. Pagnoulle, C. Detrembleur, M. Galleni and R. Jerome (2006). "New antibacterial cationic surfactants prepared by atom transfer radical polymerization." J Polym Sci Part A Polym Chem **44**(3): 1214-1224.

Li, G. J., J. R. Shen and Y. L. Zhu (2000). "A study of pyridinium-type functional polymers. III. Preparation and characterization of insoluble pyridinium-type polymers." J Appl Polym Sci **78** (3): 676-684.

Lin, F. H., Y. H. Lee, C. H. Jian, J. M. Wong, M. J. Shieh and C. Y. Wang (2002). "A study of purified montmorillonite intercalated with 5-fluorouracil as drug carrier." Biomater **23** (9): 1981-1987

Lukham, P. F. and S. Rossi (1999). "The colloidal and rheological properties of bentonite suspensions." Adv Colloid Interface Sci **82**(13): 43-92.

Ma, K. and A. C. Pierre (1999). "Colloidal behaviour of montmorillonite in the presence of Fe^{3+} ions." Colloids Surf., A Physicochem. Eng. Asp. **155**: 359-372.

Maheshwari, R. K., S. N. Sharma and N. K. Jain (1988). "Adsorption studies of phenformin hydrochloride on some clays and its application in formulating sustained release dosage forms." Indian J Pharm Sci **50**(2): 101-105.

Meng, N., N. L. Zhou, S. Q. Zhang and J. Shen (2009). "Synthesis and antimicrobial activities of polymer/montmorillonite–chlorhexidine acetate nanocomposite films." Appl Clay Sci **42:** 667-670.

Mering, J. (1946). "On the hydration of montmorillonite." Transactions of the Faraday society **42B:** 205-219.

Mondal, D., B. Bhowmick, M. D. Mollick, D. Maity, N. Saha, V. Rangarajan, D. Rana, R. Sen and D. Chattopadhyay (2014). "Antimicrobial activity and biodegradation behavior of poly(butylene adipate-co-terephthalate)/clay nanocomposites." J Appl Polym Sci **131**(7): 40079-40088.

Morton, H. E. (1983). Desinfection, sterilization and preservation. B. S. eds. Philadelphia, Lea & Febiger. **225.**

Nonaka, T., H. Li, O. Tomonari and K. J. Seiji (2003). "Synthesis of water soluble thermosensitive polymers having phosphonium groups from methacryloyloxyethyl trialkyl phosphonium chlorides–N-isopropylacrylamide copolymers and their functions." Appl Polym Sci **87** (3): 386-393.

Nonaka, T., E. Noda and S. Kurihara (2000). "Graft copolymerization of vinyl monomers bearing positive charges or episulfide groups onto loofah fibers and their antibacterial activity." J Appl Polym Sci **77**(5): 1077-1086.

Nonaka, T., Y. Uemera, K. Enishi and S. Kurihara (1996). "Antibacterial activity of resin-containing triethylenetetramine side chains and/or thiol groups–metal complexes." J Appl Polym Sci **62**(10): 1651-1659.

Nunes, C. D., P. D. Vaz, A. C. Fernandes, P. Ferreira, C. C. Romão and J. M. Calhorda (2007). "Loading and delivery of sertraline using inorganic micro and mesoporous materials." European Journal of Pharmaceutics and Biopharmaceutics **66:** 357-365.

Nzengung, V. A., E. A. Voudrias, P. Nkedi-kissa, J. M. Wampler and C. E. Weaver (1996). "Organic cosolvent effects on sorption equilibrium of hydrophobic organic chemicals by organoclays." Environ Sci and Tech **30**(1): 89-96.

Paulus, W. (2005). Directory of microbicides for protection of materials/A handbook New York Inc., springer-verlag. 157.

Pongjanyakul, T., A. Priprem and S. Puttipipatkhachorn (2005). " Influence of magnesium aluminium silicate on rheological, release and permeation characteristics of diclofenac sodium aqueous gels in-vitro." J Pharm Pharmacol **57**(4): 429-434.

Rhim, J. W., S. I. Hong, H. M. Park and P. K. W. Ng (2006). "Preparation and characterization of chitosan-based nanocomposite films with antimicrobial activity." J Agric Food Chem **54**: 5814-5882.

Salahuddin, N. and R. Abdeen (2012). "Drug release behavior and antitumor efficiency of 5-ASA loaded chitosan-layered silicate nanocomposites." Journal of inorganic and organometallic polymers and materials **23**: 1078-1088.

Salahuddin, N., S. Abo-El-Enein, A. selim and O. Salah-el-dein (2010). "Synthesis and characterization of polyurethane-urea clay nanocomposites using montmorillonite modified by oxyethylene–oxypropylene copolymer" Polym Adv Technol **21**(8): 533–542.

Salahuddin, N., M. M. Ayad and M. M. Ali (2008). "Synthesis and characterization of polyaniline-organoclay nanocomposites." J Appl Polym Sci **107**: 1981-1989.

Salahuddin, N., B. Badr and R. Abdeen (2012). "Synthesis and antimicrobial activity of biocidal polymer-montmorillonite nanocomposites." Polymers International **61**: 99-110.

Salahuddin, N., B. Badr and R. Abdeen (2012). "Synthesis, characterization and antimicrobial activities of polymer- montmorillonite/aldehydes nanocomposites." Polymer Composites **33**: 643-654.

Salahuddin, N., A. El-Barbary and N. Abdo (2009). "Delivery systems for some biologically active 1,2,4-triazine derivatives: synthesis and release." Polym Adv Technol **20**: 122-129.

Salahuddin, N., A. El-Barbary and N. Abdo (2009). "Effect of Polymethylmethacrylate-Montmorillonite Nanocomposite on the Release of some Biologically active 1,2,4-triazine derivatives." Polymer composites **30**(8): 1190-1198.

Salahuddin, N., A. El-Barbary, N. Allam and A. Hashim (2014). "Polyamide Montmorillonite Nanocomposites as a Drug Delivery System: Preparation,

Release of 1,3,4-Oxa(thia)diazoles, and Antimicrobial Activity." J Appl polym sci **131**: 41177-41190.

Salahuddin, N., E. Kenawy and R. Abdeen (2012). "Polyoxyalkylene-montmorillonite nanocomposites for drug delivery vehicle: Synthesis and characterization." J Appl Polym Sci **125**: 157-166.

Salahuddin, N., A. Moet, A. Hiltner and E. Baer (2002). "Nanoscale highly filled epoxy nanocomposite." Eur Polym J **38**: 1477-1482.

Salahuddin, N. and M. Shehata (2002). "Reduction of polymerization shrinkage in methyl methacrylate-montmorillonite composites." Materials Letters **52**: 289-294

Samour, C. M. (1976). Polymeric Drugs. L. G. Donaruma and O. Vogl. New York Academic press. 131.

Sauvet, G., S. Dupond, K. Kazmierski and J. Chojnowski (2000). "Biocidal polymers active by contact V. Synthesis of polysiloxanes with biocidal activity." J Appl Polym Sci **75** (8): 1005-1012.

Schell, T. C., M. D. Lindmann, E. T. Korngay, D. J. Blodgett and J. A. Doerr (1993). "Effectiveness of different types of clay for reducing the detrimental effects of aflatoxin-contaminated diets on performance and serum profile of weanling pigs." J Anim Sci **71**: 1226-1231.

Seema, M. D. (2013). "polymer nanocomposites as a novel drug carrier: Synthesis, characterization and controlled release study of Propranolol Hydrochloride." Appl Clay Sci **80-81**: 85-92.

Seki, Y. S. and Y. C. Kadir (2006). "Adsorption of promethazine hydrochloride with KSF montmorillonite." Adsorption **12**: 89-100.

Shemes, R., D. Goldman, M. Krepker, Y. Danin-Poleg, Y. Kashi, A. Vaxman and E. Segal (2015). "LDPE/clay/carvacrol nanocomposites with prolonged antimicrobial activity." J Appl Polym Sci **132**(2): 41261-41269.

Shetty, S. N., I. M. Khazi and C. Ahn (2010). "Synthesis, Anthelmintic and Anti-inflammatory Activities of Some Novel Imidazothiazole Sulfides and Sulfones " Bull. Korean Chem Soc **31**(8): 2337-2340.

Suci, P. A., J. D. Vrany and M. W. Mittelman (1998). "Investigation of interactions between antimicrobial agents and bacterial biofilms using attenuated total reflection Fourier transform infrared spectroscopy." Biomaterials **19**(4-5): 327-339.

Suresh, R., S. N. Borkar, V. A. Sawant, V. S. Shende and S. K. Dimble (2010). "Nanoclay drug delivery system." International Journal of Pharmaceutical Sciences and Nanotechnology **3**: 901-905.

Suzuki, N., Y. Nakamura, Y. Watanabe and Y. Kanzaki (2001). "Intercalation compounds of layered materials for drug delivery use:II. Diclofenac sodium." Chem. Pharm. Bull **49**(8): 964-968.

Takahashi, T., Y. Yamada, K. Kataoka and Y. Nagasaki (2005). "Preparation of a novel PEG-clay hybrid as a DDS material: dispersion stability and sustained release profiles." J Control Release **107**: 408-416.

Takai, K., T. Ohtsuka, Y. Senda, M. Nakao and K. Matsuoka (2002). "Antibacterial properties of antimicrobial-finished textile products." J Micobial Immunol **46**(2): 75-81.

Uchida, M. (1995). "Antimicrobial zeolite and its application." Chem Ind **46**: 48-54.

Uemura, Y., I. Moritake, S. Kurihara and T. J. Nonaka (1999). "Preparation of resins having various phosphonium groups and their adsorption and elution behaviour for anionic surfactants." J Appl polym Sci **72**(3): 371-378.

Vaia, R. A., H. Ishii and E. Giannelis (1993). "Synthesis and properties of 2dimentional nanostructures by direct intercalation of polymer melts in layered silicates." Chem Mater **5**: 1694-1696.

Wang, J., J. X. Li, L. Ren, A. S. Zhao, P. Li, Y. X. Leng, H. Sun and N. Huang (2007). "Antibacterial activity of silver surface modified polyethylene terephthalate by filtered cathodic vacuum arc method." Surf Coat Tech **201**: 6893-6896.

Wang, X., Y. Du and J. Luo (2008). "Biopolymer/montmorillonite nanocomposite: preparation, drug-controlled release property and cytotoxicity." Nanotechnology **19** (6) 1-7.

Wang, X., Y. Dua, J. Luo, J. Yang, W. Wang and J. F. Kennedy (2009). "A novel biopolymer/rectorite nanocomposite with antimicrobial activity." Carbohydrate Polym **77**: 891-897.

Wang, X., Y. M. Du, X. H. Wang, X. W. Shi and Y. Hu (2006). "Preparation, characterization and antimicrobial activity of chitosan/layered silicate nanocomposites." Polymer **47** (19): 6738- 6744.

Yano, K., A. Usuki, A. Okada, T. Kurauchi and O. Kamigaito (1993). "Synthesis and properties of polyimide–clay hybrid." J Polym Sci Part A Polym Chem **31**: 2493-2498.

Yuan, Q., J. Shah, S. Hein and R. D. K. Misra (2010). "Controlled and extended drug release behavior of chitosan-based nanoparticle carrier." Acta Biomaterialia **6**(3): 1140-1148.

Zheng, J. P., L. Luan, H. Y. Wang, L. F. Xi and K. D. Yao (2007). "Study on ibuprofen/montmorillonite intercalation composites as drug release system." Appl Clay Sci **36**: 297-301.

Zhou, Y. H., M. S. Xia, Y. Ye and C. H. Hu (2004). "Antimicrobial ability of Cu^{+2}- montmorillonite." Appl Clay Sci **27**: 215-218.